Vector-Valued Optimization Problems in Control Theory

ACADEMIC PRESS RAPID MANUSCRIPT REPRODUCTION

This is Volume 148 in
MATHEMATICS IN SCIENCE AND ENGINEERING
A Series of Monographs and Textbooks
Edited by RICHARD BELLMAN, *University of Southern California*

The complete listing of books in this series is available from the Publisher upon request.

VECTOR-VALUED OPTIMIZATION PROBLEMS IN CONTROL THEORY

M. E. SALUKVADZE
CONTROL SYSTEMS INSTITUTE
SOVIET GEORGIA
TBILISI, U.S.S.R.

Translated By

JOHN L. CASTI
DEPARTMENT OF COMPUTER APPLICATIONS
AND INFORMATION SYSTEMS
NEW YORK UNIVERSITY
NEW YORK, NEW YORK

ACADEMIC PRESS
A Subsidiary of Harcourt Brace Jovanovich, Publishers
New York London Toronto Sydney San Francisco
1979

ACADEMIC PRESS, INC.
111 Fifth Avenue, New York, New York 10003

United Kingdom Edition published by
ACADEMIC PRESS, INC. (LONDON) LTD.
24/28 Oval Road, London NW1 7DX

Library of Congress Cataloging in Publication Data

Salukvadze, Mindiia Evgen'evich.
Vector-valued optimization problems in control theory.

(Mathematics in Science and Engineering; 148)
Translation of Zadachi vektornoi optimizatsii v teorii upravleniia.
1. Control theory. 2. Automatic control.
3. Mathematical optimization. 4. Vector valued functions. I. Title. II. Series.
QA402.3.S24613 629.8'312 79-23364
ISBN 0-12-616750-8

Originally published in the U.S.S.R. under the title
Zadachi vektornoi optimizatsii v teorii upravlenia,
Metsniereba Publishing House, 1975

PRINTED IN THE UNITED STATES OF AMERICA

79 80 81 82 9 8 7 6 5 4 3 2 1

Contents

Chapter 4 Programming Optimal Trajectories for Problems with Vector-Valued Criteria

Chapter 5 A. Letov's Problem. The Analytic Construction of Optimal Regulators for Problems with Vector-Valued Criteria

Chapter 6 The Optimization of Vector Functionals in Linear (Nonlinear) Programming Problems

Chapter 7 Parameter Optimization in Engineering Systems

Introduction

The science of control began to emerge almost one and a half centuries ago. Some of the important stages in its development are discussed in [1]. To compose a survey of the basic directions in control theory, it is necessary to organize a book by an historical plan. We restrict ourselves to considering only the important role of Soviet scholars in the development and formation of this comparatively young science—on the basis of which was created high-quality and high-accuracy automatic control systems; we recommend the survey works [2,3] to the interested reader. In these works a good account is given of the contributions of Soviet scholars to the development of control theory and its extensions.

One of the most important directions in control theory is the development of methods for analyzing the quality of control processes. The fact is that contemporary progress in science and technology very sharply presents scientists and engineers with the problem of creating even more perfect automatic control systems. Up to fifty years ago the requirement on such a system was only to insure stability. More recently, the requirements have been increased and consist of the creation of higher quality systems that are ranked according to one or the other of several indices of quality. The formulation of such questions has produced an especially vital scientific field—the theory of optimal control.

Historically, the theory of optimal control processes was used for problems concerning the optimal control of aircraft maneuvers. One of the first publications in which the character of this new scientific area was exhibited was the work of D. E. Okhotsimskii [4], presenting a variant of a problem related to a ballistic rocket. In this work the regimes of mass expenditure for which some characteristics of the motion are extremal were studied. In speaking about the formation of optimal control theory, we must note the great service of A. A. Fel'dbaum in posing the general problem of minimal time [5] and in studying several of its particular cases.

Fruitful development of optimal control theory began in the 1950s. New mathematical methods appeared for the calculation of optimal controls. In this connection we note the maximum principle of L. S. Pontryagin [6,7], Bellman's method of dynamic programming [8,9], methods of functional analysis [10–17], the method of analytic construction of optimal regulators [18], methods of operations research [19–27], and so on. Later the theory of optimal processes was developed for application to discrete systems [28–32], and there appeared numerical methods for computing optimal controls based upon the ideas of successive analysis of variants and dynamic programming [33].

The fundamental results in the theory of optimal control, obtained by L. S. Pontryagin, N. N. Krasovskii, and R. Bellman, were extensively developed in the works of the Soviet scholars V. G. Boltyanskii, A. G. Butkovskii, R. Gabasov, R. V. Gamkrelidze, Yu. V. Germeier, A. Ya. Dubovitskii, V. I. Zubov, Yu. N. Ivanov, F. M. Kirillova, A. A. Krasovskii, V. G. Krotov, A. M. Letov, A. I. Lur'e, A. A. Miliutin, E. F. Mischenko, N. N. Moiseev, A. A. Pervozvanskii, B. N. Pshennichnii, B. S. Razumikin, L. I. Rozonoer, V. V. Tokarev, V. A. Troitskii, A. A. Fel'dbaum, G. L. Kharatishvili, Ya. Z. Tsypkin and the foreign scholars M. Athans, A. Balakrishnan, L. Zadeh, R. Kalman, R. Kulikowski, J. Kurzweil, J. LaSalle, G. Leitmann, A. Miele, L. Neustadt, S. Chang, and many others.

Optimal control theory deals with a set of questions that are surveyed in [34–36] and also in [6–32]; more recent works are treated in [43].

We note that in all of the above-mentioned publications the problems are considered from the viewpoint of the optimization of a fixed scalar functional given in advance. However, in many practical problems it is not sufficient to consider the optimization of control systems with a single scalar criterion [39,47]. Therefore, further development of control methods for problems with vector-valued criteria is needed.

In one of the new areas of optimal control theory, the theory of differential games, problems of the pursuit of one controlled object by another are studied; in these problems features appear pertaining to the attainability of different, conflicting objectives. In problems of pursuit the partners tend to select strategies in order that the chosen criteria give different values, which are in conflict. In other, more general problems in differential games, the partners have their own indices of quality that are different, but depend upon the strategy of each player; the players tend to choose strategies that are best in the sense of their own indices of quality [11,12,20].

The peculiarity of differential games rests in the fact that for optimization of each index of quality there is a characteristic control resource. In control theory the situation is different. The rule of consumption of control resource is chosen so that all performance indices of the system take on their best values. However, the selection of a control law is effected by optimizing only one of the chosen criteria, which, as noted above, is not sufficient in many practical situations.

The lack of control rules that are well defined through conditions of optimizing a scalar criterion leads to the following: The selection and optimization of a given functional answers to only one of the requirements of the control system, while other requirements, which are often equally important, are ignored. The totality of all requirements may be accounted for by a collection of functionals forming a vector of criterion functions, i.e., a vector-valued criterion. The formulation of requirements for automatic systems in the form of a set of optimality criteria apparently must be empirically arrived at, after which a definite control rule may be determined by application of mathematical techniques [6–33].

Very naturally there arise mathematical problems requiring the simultaneous optimization of a collection of functionals, each of which measures a definite aspect of the system. Important problems of simultaneous optimization of two or more functionals are repeatedly noted by Soviet and foreign scholars [37,47,48].

The current work is devoted exactly to this problem. In this work we study problems of programming optimal trajectories, the analytic construction of regulators, and linear and nonlinear programming under vector-valued criteria. For the solution of all these problems, we employ a single approach based on the definition of an ideal (utopian) point and on the introduction of a norm into the space of optimizing functionals.

Naturally, the choice of a norm in the space of optimizing functionals determines a definite solution of the problem. We study the above-mentioned problems for the particular case in which we choose an euclidean space of functionals. Such a step, i.e., the choice of a quadratic measure, results in a situation that in many practical problems leads to exceptionally good results, and also we obtain a more transparent form of the mathematical presentation. The influence of the measure on the solution of the problem must be studied separately for each concrete problem. Generally, it is reasonable to choose a quadratic metric and to study the question of the stability of the solution to the problem under variation of the measure.

In the first chapter of this monograph we give a survey of the existing literature on the problem of vector optimization and more explicitly consider a number of works that are similar in theme. We give an assessment of each of these works.

In the second chapter we formulate the general mathematical statement of the problem of optimizing vector functionals, based on the idea of defining an ideal point in a space of optimizing functionals and approximating the values of the quality indices of the system to this point. We discuss the expediency of such an approach to the solution of the problem using a concrete example.

The third chapter is devoted to a discussion of the problem of the existence of a solution to the problem of optimizing vector functionals presented in the spirit of Chapter 2. We introduce necessary conditions for the existence of such a solution in problems of programming optimal trajectories. These conditions are illustrated with the concrete example of flight for an autopilot aircraft.

In Chapter 4, on the basis of the Pontryagin maximum principle formulated for the problem of Mayer, we solve the general problem of programming optimal trajectories for vector-valued criteria. To illustrate the method we solve the problem of optimal flight of a rocket to a given point.

In Chapter 5 we discuss the problem of analytic construction of optimal regulators for vector-valued criteria. Along with the general problem, we give a solution to the linear problem of analytic construction for vectors whose components are integrals of quadratic indices.

Chapter 6 is devoted to problems of operations research for vector goal functions. The general problems of linear and nonlinear programming are considered, questions of existence of a solution under vector goals are studied, compromise solutions are discussed, and concrete examples of planning for metallurgical operations are solved.

In Chapter 7 of the monograph, we solve a practical problem from the domain of nuclear reactor installations. With the help of the above methods we solve a problem of determining optimal parameters of the heating machinery of an atomic power station from the point of view of minimizing both weight and volume of condensers, as well as the cost of cooling the system. The solution consists of a number of results. The results of this chapter were obtained together with A. N. Yoseliani, A. A. Michailovizh, and V. B. Nesterenko.

CHAPTER I
A SURVEY OF OPTIMIZATION PROBLEMS WITH VECTOR CRITERIA

1. General Survey of the Problem

The problem of optimization of a vector-valued criterion arises in connection with the solution of problems in the areas of planning and organization of production. However, recently it has rapidly spread to dynamical control systems. Currently, the problem of optimizing vector-valued criteria has become a central part of control theory and great attention is given to it in the design and construction of modern automatic control systems.

A little history.

The first formulation of a problem of optimizing vector-valued criteria was given in 1896 in a publication of the economist Pareto (49). More than a half a century later questions of multicriteria optimization were touched upon in the publications (50,51), but essential developments of the problem were not carried out.

The fundamental work of von Neumann and Morgenstern (19) on game theory and economic behavior, published in 1944, laid the basis for the concept of a solution in the case of several conflicting criteria. An account of these ideas is given in (21).

In 1963 Professor Lotfi Zadeh published the article (52), which first presented the question of designing control systems which are optimal relative to several performance indices. In spite of the fact that the paper did not contain a rigorous mathematical solution of the general problem of vector optimization, the work must be reckoned as the beginning of a new, more global concept of optimization.

Next, as a continuation of the arguments of (52), in the article (53), a concrete example, using simple linear programming, was given to show the behavior of systems subject to optimization with respect to different objective functions.

In 1964, V. Nelson (48) formulated the general multicriteria optimal control problem for dynamical systems. However, he reduced the problem to one in which only one of the scalar criteria was optimized under isoperimetric constraints on the values of the other functionals. Here already is expressed the idea of optimization of a functional of the solution of a problem which is obtained as the solution of an optimization problem using other functionals not connected with the first. In this work, a simplified aircraft problem is considered for the minimization of fuel expenditure under a flight-time constraint.

In 1966 Professor Sheldon Chang, in his general theory of optimal processes (54) developed necessary conditions for so-called non-improvable solutions in dynamical problems of multicriterion optimization. At the same time, a series of works (55-58) were presented dealing with the determination of necessary conditions for the existence of solutions for multicriteria optimization problems in finite-dimensional and in linear topological spaces. Here, the distinctive approach of "scalarization" was used, i.e. the reduction of the vector optimization problem to a family of optimization problems with a scalar performance index. By scalarization is meant the representation of a vector criterion by a linear combination of its components with strictly positive (real) coefficients.

In the work (59), published in 1967, the ideas of Nelson (48) about optimization of a system with respect to a given scalar criterion were developed, determining another scalar performance index from the optimization conditions. Such an approach obtained the name optimization of an ordered sequence of criteria.

In the same year, Frederick Waltz published the article (60) which in agreement with the ideas of L. Zadeh on the problem of optimization of vector-valued functionals, presented a method for its solution based upon the principle of hierarchical ranking. The principle consists of the following: determination of a numerical matrix. Each element of this matrix Φ_{ij} defines a number which is obtained by calculating the value of the the jth functional. Next, by estimating the gain and loss of different performance indices, we choose the principal criterion of the system and, by the same token, we establish a hierarchical subordination of the criteria.

The problem of multicriteria optimization was widely developed in the 1970s. Here we note both the work of foreign scholars (61-101) and the local work (102-135).

The problem of vector optimization in simplified form consists of the following: the behavior of a system is characterized by an n-dimensional vector $x = \{x_1,\ldots,x_n\}$, $x \in X \subset E^n$ and is measured by a k-dimensional vector function $I(x) = \{I_1(x),\ldots,I_k(x)\}$, the components of which represent given real functions of x. It is required to determine a point $x^0 \in X$, optimizing in some sense the values of the functions $I_1(x),\ldots,I_k(x)$.

The solution methods in the above-mentioned works for the problem of vector optimization may be classified into the following groups:

1) optimization of a hierarchical sequence of performance indices,

2) determination of a set of unimprovable points,

3) determination of a solution based upon one form of compromise or another.

The method of <u>optimization of a heirarchical sequence of performance criteria</u> is based upon the introduction of a preference ordering of the given criteria so that the preference order is reduced to an ordering of scalar criteria. After the establishment of such an order, assumed to be $I_1(x), I_2(x),\ldots,I_k(x)$, the solution $x^0 \in X$ is determined as that point satisfying the relations

$$I_1(x^0) = \min_{x \in X_0 \subset X} I_1(x),$$

$$I_2(x^0) = \min_{x \in X_1 \subset X_0} I_2(x),$$

$$I_3(x^0) = \min_{x \in X_2 \subset X_1} I_3(x),$$

$$\cdots\cdots\cdots\cdots$$

$$I_k(x^0) = \min_{x \in X_{k-1} \subset X_{k-2}} I_k(x).$$

Here the set $X_i \subset X_{i-1}$ $(i=1,2,\ldots,k-1)$ is defined as

$$X_i = \{x : I_i(x) = \min_{x \in X_{i-1}} I_i(x)\}.$$

As was noted, the idea of such an approach to the solution of vector optimization problems was proposed in (48). The works (59, 60,102) were carried out in this direction and we shall devote special attention to them later. Here we note only that in (60), on the basis of engineering considerations, the principal criterion was established while the remaining criteria were declared hierarchically subordinate to it.

Application of the given method becomes less effective for the solution of the majority of practical problems since optimization with respect to the first, most important criterion already leads to a unique optimal solution and everything is reduced to optimization only relative to the first criterion.

Several numerical problems connected with the choice of the optimal structure of hierarchical control systems are discussed in (103).

The largest collection of work is devoted to the <u>method of determination of the set of unimprovable points</u>. A point $x^0 \in X$ is called unimprovable in X relative to I(x), if among all $x \in X$ there does not exist a point $\bar{x}$ such that $I_\alpha(\bar{x}) \leq I_\alpha(x^0)$, $\alpha = 1,2,\ldots,k$, with at least one of the inequalities being strict. Here, at first we note the work (52), in which the definition of an unimprovable point is first given, together with a discussion of the problem. In (61) several properties of the set of unimprovable points are studied in the general problem of dynamical systems with k performance criteria and it is shown that the problem reduces to the minimization of a linear form of the components of the vector I(x) with constant weighting coefficients, i.e. to the minimization of the expression

$$I_\lambda = \sum_{\alpha=1}^{k} \lambda_\alpha I_\alpha(x),$$

in which $\lambda_\alpha > 0$, $\alpha = 1,2,\dots,k$, $\sum_{\alpha=1}^{k} \lambda_\alpha = 1$.

Under the name "The General Theorem of Optimal Control", in (54) a theorem on necessary conditions for unimprovable solutions in the general optimal control problem with vector-valued criteria was given. Later, we shall formulate this theorem. The results of (54) were generalized to linear topological spaces in (55-58). Necessary and sufficient conditions for the optimality of a control for differentiable criteria functions were given in (62,63).

Necessary and sufficient conditions for the existence of an unimprovable solution in mathematical programming problems and the separation of dominant solutions from the set of unimprovable points are dealt with in the series of works (65-68).

From the important works dealing with operations research problems, we should isolate works in which the problem of admissibility of solutions in complex situations are studied. One of the characterisitc features in the theory of admissible solutions is the presence of a large number of criteria of the utility of the system and the impossibility of comparing these criteria because of the complexity of the system. Several questions connected with the relative dependence of the criteria are studied in (69).

Since the admissibility of a solution must be realized on the basis of a comparison of a set of alternatives, i.e. a choice of points from a set of unimprovable points, in many practical situations it is expedient to employ man-machine procedures. The idea of using man to effect the solution in one or another form was expressed in (70). Already in this paper a method was proposed based upon obtaining information from humans at several levels in the solution process and using this information at higher levels.

The process is: presentation of intermediate information to the person about the solution, machine computation using additional information supplied by the person and repetition of this cycle until an acceptable best-possible solution is obtained. Exhaustive surveys of the man-machine procedure for acceptance of a solution is given in (104,105).

A method allowing the consideration of incommensurable criteria was presented in (71). The idea of the method consists in a detection of successively less-disputed relations between the alternatives and in a declaration of their incommensurability.

With the goal of obtaining information needed for justifying an admissible solution, in (106,107) it was proposed to effect a decomposition of the set of states into subsets for each of which an alternative solution is optimal. The role of the measure of information in problems of admissible solutions is discussed in (108-110).

Admissible solutions in complex situations are also dealt with in the works (111-119). In the works (120-124) are studied some problems arising in the admissibility of solutions in the control of large systems. Exhaustive surveys on the basic problems and methods of vector optimization are given in (125,126).

Methods of physical modeling in mathematical programming and economics are presented in (127,128). The effective determination of the equilibria of linear economic models, consisting of economic systems having characteristic goals and budgets, has been carried out using these methods. A very important result of these investigations is the establishment of the existence of a global goal function, having the physical meaning of system entropy, and the determination of recurrence relations for the optimal distribution of resources between systems.

In game theory, the method of unimprovable points is called a Pareto-optimal solution. Currently, the term "Pareto-optimal" is widely used in the theory of vector optimization. The properties of Pareto-optimal solutions in cooperative games are studied in (72-87) and necessary and sufficient conditions for the existence of such a solution are given.

The connection between the method of optimization of a hierarchical sequence of criteria and the method of determination of unimprovable points is investigated in (129). Cases are established when the solution $x^0 \varepsilon X$, optimizing a hierchical sequence of criteria, is an unimprovable point in X with respect to the same criteria.

In (130) is studied the effect of the method of curtailment of the vector criterion on the form of the optimal solution in vector optimization problems and it is shown that the optimal trajectory always belongs to a single class of curves, independent of the method of curtailment. A discrete stochastic system with two criterion functions is studied in (131).

Methods for determining a solution based on one form of compromise or another have recently been increasingly often applied for the solution of wide classes of vector optimization problems. The viewpoint that a choice of a solution in the face of several criteria must be confined only to the identification of a compromise region, while identification of a unique optimal solution in a definite sense of solution is either impossible to carry out or is carried out on the basis of a random choice or on the basis of heuristic considerations, has been convincingly refuted in recent years. In the works (132-135), the principle of equitable compromise for choosing a solution in vector optimization problems has been presented. The principle is based upon the idea that the relative decrease in value of one or several criteria should not exceed the relative increase in the remaining criteria. In (136-138), an unconditional preference criterion is used for the synthesis of a system with a vector criteria.

One of the possible compromise approaches is to consider a preliminary choice of weighting coefficients and to define an optimal solution as that which minimizes the sum $\lambda_1 I_1(x) + \ldots + \lambda_k I_k(x)$, in which $\lambda_1, \ldots, \lambda_k$ are the already chosen numbers. The choice of coefficients

$$\lambda_1 = \frac{1}{I_{10}}, \quad \lambda_2 = \frac{1}{I_{20}}, \ldots, \lambda_k = \frac{1}{I_{k0}},$$

has been used in (139), where $I_\alpha(x)$ obtained by optimizing only with respect to itself. This idea was used by Ch. Yuttler (140) to study linear economic models with several conflicting goal functions. Such an idea is contained in the work (141), in which different approaches to aggregating part of the criteria into a single criterion are considered.

In (142-149), a new approach to the solution of multicriteria optimization is presented based upon the idea of an ideal (utopian) point in the space of criteria and a norm is introduced into this space. The compromise solution obtained under this method is a Pareto-optimal solution and ensures the closest approach of the criteria to their highest values. The state approach is free from the choice of weighting coefficients $\lambda_1, \ldots, \lambda_k$ which is an especially difficult matter for the solution of control problems in large systems.

Similar ideas were expressed almost simultaneously in the work (89) on dynamical vector optimization problems and in (124) on statistical vector optimization. More details on the work of (89) are given below. The same idea was applied in (127) for the solution of the problem of optimal design of universal systems.

The above-mentioned approach for the case of statistical systems was developed in (91-99). In these works, dominant solutions were isolated from Pareto-optimal solutions defining an approach to minimization through a norm more general than a quadratic and the best possible solution was chosen from this dominant set on the basis of additional arguments and information obtained from the human decisionmaker, leading to the final solution. Several results from these works are discussed later in Chapters IV and VI.

We note that the choice of the measure of approximation to the ideal point in the space of criteria, i.e. the definition of the metric in this space, strongly influences the form of the final solution to the vector optimization problem. Consequently, the question of the choice of such a measure must be carefully studied for each concrete problem. However, in a series of practical cases the choice of a quadratic metric as a measure of approximation to the ideal point in euclidean space gives effective results, as is illustrated in Chapter VII, where we use it to determine the optimal parameters of a boiler in a nuclear power station and it is also used in (124) for the solution of the problem of controlling a chemical engineering process under a wide choice of criteria.

An axiomatic approach, based upon the establishment of requirements for the terminal solution in each concrete vector optimization problem is presented in (116), and this information is used for determining an approximation measure to the ideal values of the criterion functions. The axiomatic approach considerably reduces the indefiniteness in the problem. However, it is not easy to check the conditions in concrete situations and, moreover, it is difficult to formulate conditions which, on the one hand, are "reasonable" and, on the other hand, significantly reduce the class of possible solutions.

It seems natural to consider methods having the ability to reduce the indefiniteness at the expense of the time needed to obtain information about the problem, which permits us to obtain the necessary results.

Here we note the "method of constraints" (180) which consists of the following: one starts with a solution, relative to which it is necessary to impose constraints on particular criteria, which must be satisfied in the first place. On the basis of this informations, a revision of the solution is affected.

In (116) an adaptive algorithm is presented for cases when we have a preference ordering among the criteria. Under this procedure, there also takes place an extension of the solution process with the goal of progressively adding information about the problem under study to ensure the determination of a satisfactory solution.

In recent studies of A.I. Kaplinskii, A.S. Krasnenkeva and Ya.Z. Tsypkin, methods of randomization and smoothing are investigated in the context of adaptive algorithms. The methods are applied in those cases when it is necessary to repeatedly obtain a solution under conditions of random noise and when the particular criteria are not formally expressed.

An interesting new idea, based upon obtaining a solution in a fuzzy environment, is presented in (100) and is considered further in section 7 of this chapter.

Now we discuss several works in more detail.

2. The Remarks of L.A. Zadeh

The discovery of the maximum principle (6,7), the development of dynamic programming (8-11) and the methods of functional analysis (12-17), made the theory of optimal control systems with a scalar criterion almost complete up until the publication of the remarks of L.A. Zadeh in (52). In this work, the author first turned the attention of specialists to the fact that in a majority of cases of practical concern, we "naturally" wish to design an automatic control system which has many optimal characteristics, not one or the other.

As noted by A.M. Letov (37) in a survey paper presented to the 2nd IFAC Congress, the "naturalness" of this desire is of exactly the same character as the natural wish of Agaf'i Tikhonov to have one fiancé having all the best features from the set of all fiances proposed to him (N.V. Gogol, "Marriage").

We almost always want most machines and structures to have multicriteria optimal properties. The scientific goal of L.A. Zadeh's paper, in addition to its initiatives, is to show that in the majority of practical cases, exact optimization of a vector functional is unattainable. More precisely, this means that if a control is chosen to optimize one of the scalar functionals, then almost always it is not possible to optimize a second scalar functional using the same control.

As we show later, the acceptance of this conclusion coming out of Zadeh's work is a critical aspect in understanding problems of vector optimization. Its value, in essence, is the indirect assertion that the problem of exact optimization, of vector criteria, i.e. the problem of optimizing all scalar criteria with the same control, is impossible and to give it a precise mathematical formulation is also impossible. The main point is the inconsistency of the natural goal of multicriteria optimization and the impossibility of satisfying this goal by exact methods.

In L. Zadeh's remarks there arises the idea of choosing a best possible solution from the set of unimprovable solutions on the basis of additional considerations.

3. Optimization of an Ordered Set of Scalar Criteria

We now state an optimization problem satisfying a set of scalar criteria.

We assume that we are given a mathematical model of the object to be controlled. This means that we are given

1) the equations of motion in vector form

$$\dot{x} = F(x,u,t), \tag{1.1}$$

in which $x \in R^n$, $u \in R^m$;

2) the region of definition

$$N(x,u,t) \geq 0; \tag{1.2}$$

3) a time interval $\tau = (0,T)$ or $\tau = [0,\infty)$;

4) a set V of piecewise-continuous functions taking values in R^m, called the set of admissible controls;

5) a function $F(x,u,t)$, which is smooth in x,u in the region (1.2).

We assume that we are given boundary conditions of the problem, which we symbolically represent by the equation

$$(i,f) = 0. \tag{1.3}$$

Also, we assume that we are given a vector function

$$I(u) = \Phi(x,u,t) \tag{1.4}$$

with components

$$I_a(u) = \Phi_a(x,u,t), \quad (\alpha=1,\ldots,k). \tag{1.5}$$

Along any admissible trajectory x(t), u(t), each function $I_\alpha(u)$ assumes a definite numerical value and, consequently, I(u) represents a vector functional.

We assume that $\Phi_1(x,u,t),\ldots,\Phi_k(x,u,t)$ are smooth functions in (1.2). We will say that the set (1.5) of scalar criteria is ordered with respect to the index $\alpha = 1,2,\ldots,k$ if there exist subsets $V_1 \subset V, V_2 \subset V_1, V_3 \subset V_2, \ldots, V_k \subset V_{k-1}$ of admissible controls on which the lower bounds of all the scalar functionals are assumed, i.e.

$$\Delta\Phi_1 = \min_{\forall u \in V_1} \Phi_1(x,u,t)\big|_\tau,$$

$$\Delta\Phi_2 = \min_{\forall u \in V_2} \Phi_2(x,u,t)\big|_\tau, \tag{1.6}$$

$$\cdots\cdots\cdots\cdots$$

$$\Delta\Phi_k = \min_{\forall u \in V_k} \Phi_k(x,u,t)\big|_\tau.$$

From this definition clearly follows its restricted character, the specification of which is contained in the following: we will optimize with respect to each scalar functional, in turn, using the maximum principle (6). Then if the set V_1 does not consist of a single element and $\phi_1(x,u,t)$ assumes its lower bound $\Delta\phi_1$, on each element then it is possible to optimize the functional $\phi_2(x,u,t)$ over this set. Let V_2 be the set of elements on which the functional $\Phi_2(x,u,t)$ assumes its lower bound $\Delta\Phi_2$. In this case, it is possible to optimize the functional $\Phi_3(x,u,t)$ over V_2 and so forth.

Obviously, in the case of an ordered set of scalar criteria (1.5), the set V_k will not be empty.

An account of the idea for the solution of vector optimization problems was described in (48).

A simple example of such a problem is considered in (59). Namely, we consider the vector equation

$$\dot{x} = A(t)x + B(t)u, \quad x(t_0) = x_0, \tag{1.7}$$

together with the two scalar equations

$$\dot{x}^{(1)} = f^{(1)}(x(t),t) + h^{(1)}(u(t),t), \quad x^{(1)}(t_0) = 0,$$

$$\dot{x}^{(2)} = f^{(2)}(x(t),t) + h^{(2)}(u(t),t), \quad x^{(2)}(t_0) = 0, \tag{1.8}$$

where $A(t)$ is $n \times n$, $B(t)$ is $n \times m$, the matrices are continuous on τ, $f^{(1)}(x,t) \in C^1$, $f^{(2)}(x,t) \in C^1$, $h^{(1)}(u,t) \in C^0$, $h^{(2)}(u,t) \in C^0$ are real convex functions.

Consider the set $V(t_0,T)$ of measurable functions defined on the interval τ; the domain (1.2) is convex in u and does not depend upon x, t.

For the set of scalar functionals, we take the set consisting of

$$\begin{aligned} I_1(u) &= \Phi_1(x^{(1)}) = x^{(1)}(T), \\ I_2(u) &= \Phi_2(x^{(2)}) = x^{(2)}(T). \end{aligned} \tag{1.9}$$

The problem consists in finding an admissible control $u_0(t)$ under which

$$\begin{aligned} I_1(u^0) &\leq I_1(u), \\ I_2(u^0) &\leq I_2(u) \end{aligned} \tag{1.10}$$

for all $u \in V$ and an admissible trajectory $x^0(t,u^0(t))$ satisfying the boundary condition (1.3). The control $u^0(t)$ is called optimal with respect to the vector functional.

We assume that the solution of this problem exists. Then its geometrical interpretation takes on the following form: we consider the space R^{n+2} consisting of those x, $x^{(1)} \geq 0$, $x^{(2)} \geq 0$ (Fig. 1). Let $G(x,x^{(1)},x^{(2)})$ be the set of attainable systems (1.7), (1.8) and let the point $f \in G$. Further, let G', G'' be the orthogonal projections of the set G onto the subspace R^{n+1} of x, $x^{(1)}$ and R^{n+1} of x, $x^{(2)}$, respectively. Then it is obvious that if $\underline{ief}$ is an optimal trajectory in the space R^{n+2} corresponging to the control $u^0(t)$, its projections $\underline{ie'f'}$, $\underline{ie''f''}$ must also be the optimal trajectories in the subspaces $R^{n+1}(x,x^{(1)})$ and $R^{n+1}(x,x^{(2)})$, respectively. In this event $u^0(t) \in V_2 \subset V_1 \subset V$. Obviously, the conditions of the problem must be such that the trajectory $\underline{ie'f'}$ is an

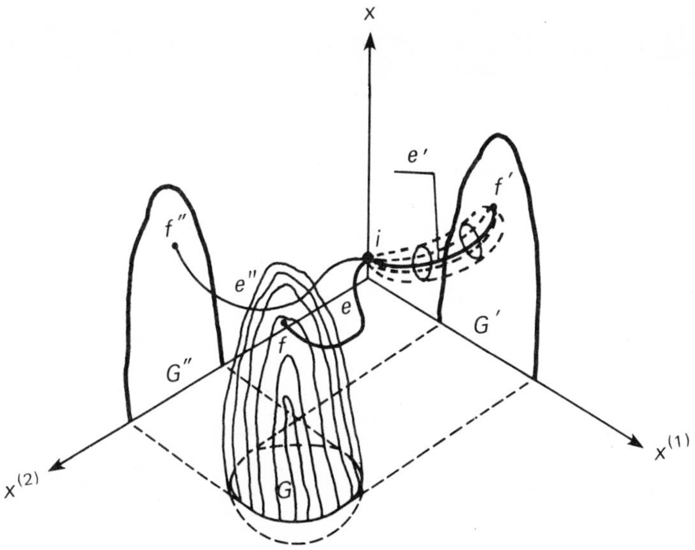

Fig. 1.

orthogonal projection of the trajectory <u>ief</u> and lies in the set of optimal trajectories in the subspace $R^{n+1}(x,x^{(1)})$ obtained for $u \in V_1$.

What these conditions are is unknown. However, it is <u>necessary</u> that the set V_1 not consist only of a single element. This is a very stringent condition.

We turn to a simple example of A. Miele (18) (cf. 24-26).

We consider the equations of an aircraft at a given altitude

$$\dot{x} = v,$$

$$\dot{v} = \frac{c\beta - Q(m,v)}{m}, \qquad (1.11)$$

$$\dot{m} = -\beta, \quad 0 \leq \beta \leq \overline{\beta}, \quad \tau = [0,T].$$

Here x is the distance travelled, v is the velocity of the aircraft, m is its mass, β is fuel consumption, c, $\overline{\beta}$ are given positive numbers, Q(m,v) is a known function characterizing the force of aerodynamic drag and T is free. We consider the plane of the variables m, v (Fig. 2). As was shown in (18), there exists a set V_1 of piecewise-continuous controls for which:

1) the condition (1.3) is satisfied, i.e. $x_i = 0$, $v_i = v_0$, $m_i = m_0$, $v_f = v_T$, $m_f = m_T$;

2) the distance travelled assumes a maximum $x_f = x_T$.

The set V_1 satisfies the maximum principle and consists of an uncountable set of different combinations of $\beta = 0$, $\beta = \bar{\beta}$ corresponding to the part of the trajectory situated on the curve $S(m,v) = 0$ (Fig. 2). The curve $S(m,v) = 0$ is known. It appears that the existence of the set V_1 gives the possibility to formulate, in principle, the vector optimization problem (6,59). However, A. Miele showed that this is actually not the case as the set V_1, under the Legendre-Clebsch conditions, consists of the single element impnf (Fig. 2) and not an uncountable number of elements (18).

As shown by the above example, even in simple cases of ordered optimization of a sequence of criteria, it may not be possible to formulate an exact statement of the problem.

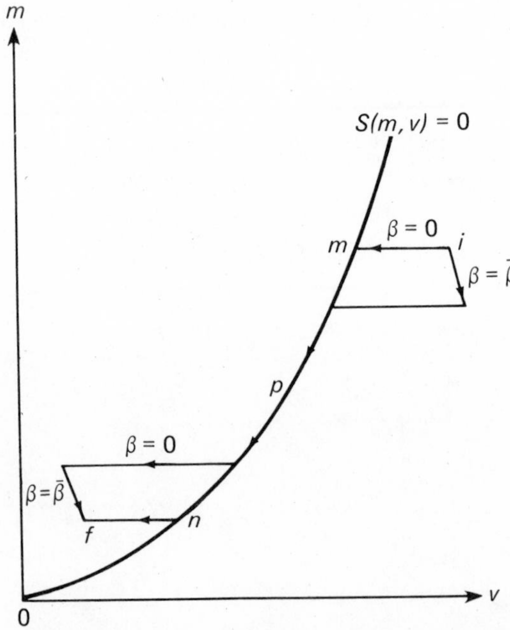

Fig. 2

Optimization problems involving an ordered collection of criteria are frequently encountered in various areas of applied mathematics, operations research and also in mathematical economics.

Similar problems are mathematically formulated in the following manner: we are given a set of admissible solutions V and an ordered collection of criteria (objective functions) $\{F_1(u), F_2(u), \ldots, F_m(u)\}$, $u \in V$. The sets

$$V_k^* = \{u \mid F_k(u) = \sup_{v \in V_{k-1}^*} F_k(v), u \in V_{k-1}^*\}, \qquad (k=1,\ldots,m). \tag{1.12}$$

are recursively defined, where $V_0^* = V$. It is required to find the set V_m^*. In (102) it is shown that a solution of the problem exists, i.e. the set V_m^* is not empty, if V is compact and all the functionals $F_k(u)$ are upper-semicontinuous.

4. Hierarchical Ranking of Scalar Criteria

The problem of hierarchical ranking of scalar criteria is studied in (60). In accordance with the ideas of L. Zadeh on optimization process with vector functions, the author presents a solution method whose basis lies in the principle of hierarchical ranking. This principle takes the following form.

Let there be given the mathematical model (1.1)-(1.3) of the object to be controlled and the vector criterion (1.4). We consider its first scalar component $\Phi_1(x,u,t)$. Let the control $u^{(1)}$ minimize this functional and assume Φ_1 takes on the value $\Phi_1^{(1)} = \Phi_1(x,u^{(1)}t)$. We compute the values of the remaining functionals using $u^{(1)}$ as $\Phi_2^{(1)}$, $\Phi_3^{(1)}$, and so on.

Now we consider the criterion $\Phi_2(x,u,t)$. Let the control $u^{(2)}$ provide the minimum of Φ_2 as $\Phi_2^{(2)}$, i.e. $\Phi_2^{(2)} = \Phi_2(x,u^{(2)},t)$. We calculate the values of the remaining criteria using $u^{(2)}$ as $\Phi_1^{(2)}$, $\Phi_3^{(2)}$, etc.

We repeat the above operation for all of the scalar functionals. By means of the above steps, we obtain the matrix

$$\begin{Vmatrix} \Phi_1^{(1)}, & \Phi_2^{(1)}, \ldots, \Phi_k^{(1)} \\ \Phi_1^{(2)}, & \Phi_2^{(2)}, \ldots, \Phi_k^{(2)} \\ \cdots & \cdots \\ \Phi_1^{(k)}, & \Phi_2^{(k)}, \ldots, \Phi_k^{(k)} \end{Vmatrix} \tag{1.13}$$

which we call the characteristic matrix. Since the matrix is finite, analysis of the numerical values of its elements estimating the gains (and losses) of different choices of system criteria allows us to make a definite choice of the principal criteria. In this case, the remaining criteria will be hierarchically subordinate among themselves.

This idea is pursued in (60) for the system

$$\dot{x}_1 = x_2, \quad \dot{x}_2 = u,$$

for the boundary conditions

$$x_1(0) = 0, \quad x_1(1) = 20,$$

$$x_2(0) = 0, \quad x_2(1) = 0$$

and the following components of the criterion vector

$$\Phi_1 = \min_u \sup_t |u(t)|,$$

$$\Phi_2 = \min_u \int_0^1 u^{(2)}(t)\,dt,$$

$$\Phi_3 = \min_u \sup_t |x(t)|.$$

In this case, the characteristic matrix (1.13) has the form

$$\begin{Vmatrix} 80 & 6400 & 40 \\ 90 & 4982.4 & 32.01 \\ 90 & 5400 & 30 \end{Vmatrix} \tag{1.14}$$

The values of the elements of the matrix (1.14) show that increasing the peak amplitude of the control u(t) from 80 to 90 (i.e. by 12.5%), the gain in energy expenditure is 15.4% and simultaneously reduces the peak amplitude of the velocity $x_2(t)$ by 25%. Hence, hierarchical optimization implies that we prefer the criterion Φ_2.

The hierarchical ranking idea presented here is attractive in its simplicity for application in simple cases for the solution of the problem of programming optimal trajectories. It may be practically applied in all cases in which the characteristic matrix indicates a criterion highly sensitive to the control and indicates a clear preference for one of the criteria from the others. Such a preference allows us to determine the hierarchical subordination of the criteria. The difficulty in the choice of a preferred criterion remains in all other cases in which the optimizing criterion is weakly sensitive relative to the chosen control and the elements of the matrix entering into the same column do not undergo important changes with changes in the superscript index, i.e. with changes in the control function.

5. Unimprovable Solutions in the Problem of Vector Optimization (Pareto Optimality)

The definition of an unimprovable system was first put forth by L. Zadeh (52). Let C denote a subclass in Σ defined by the set of values of a system S. Now we introduce the following notation:

1) $\Sigma_{>}(S)$ is the collection of all systems each of which is better than the system S;

2) $\Sigma_{\geq}(S)$ is the collection of all systems each of which is at least as good as the system S;

3) $\Sigma_{\sim}(S)$ is the collection of all systems which cannot be compared with S.

It is clear that each system from Σ lies in one of the collections defined above and the sets $\Sigma_{>}(S)$, $\Sigma_{\geq}(S)$ and $\Sigma_{\sim}(S)$ determine Σ.

Definition. A system S_0 is unimprovable in C if the intersection of the sets C and $\Sigma_{>}(S_0)$ is empty. In other words, this means that there is no other system in C which will be better than the system S_0.

The above definition includes as a special case those situations when Σ is determined by a vector criterion. For example, let S be determined by the values of a vector $x\{x_1, x_2, \ldots, x_n\}$ and let $C \varepsilon R^n$. Further, let the index of quality of the system S be measured by the components of a vector $p(x)\{p_1(x), \ldots, p_n(x)\}$ in which each $p_i(x)$ is a real function of x.

A system S_0, defined by the value of a vector $x_0 \varepsilon R^n$, is unimprovable in R^n if for any $x \varepsilon R^n$ we have

$$p_i(x_0) \leq p_i(x) \quad (i=1, \ldots, m).$$

(We are considering the case of minimization of all indices of quality $p_i(x)$.)

Sheldon Chang's control problem was formulated in the following way (54). The controlled object is described by the differential equations

$$\dot{x}_i = f(x,u,t) \quad (i=1, \ldots, n)$$

where $x\{x_1,\ldots,x_n\}$ is the system state vector, and $u\{u_1,\ldots,u_m\}$ is the control vector. The functions $f_1(x,u,t),\ldots,f_n(x,u,t)$ are continuous in x,u and continuously differentiable in x. The given data $x(t_1)$ and the time interval $T = \{t:t_1 \le t \le t_2\}$ completely determine x(t) for a given u(t). It is required that the control function u(t) satisfy the following three conditions:

a) u(t) is contained in the class of operatively* convex functions C on the interval T;

b) the trajectory x(t) corresponding to the control u(t) must remain in the admissible region X, i.e. $x(t) \in X$ for all $t \in T$;

c) the trajectory x(t) must assume the given value $x(t_2)$, where t_2 may be fixed or free, satisfying $t_2 \le T$.

A control satisfying a) and b) is called tolerable. A control which also satisfies c) is called admissible.

The performance index of the system is given as

$$y_i(t) = \int_{t_1}^{t_2} g_i(x,u,t)dt \qquad (i=1,\ldots,N).$$

---

*A set of functions C is operatively convex if it has the following property: if any function u(t) and its infinitesimal variations δu_1 and δu_2 are such that $u \in C$, $u + \delta u_1 \in C$, $u + \delta u_2 \in C$, then for any h in the interval $0 < h < 1$, we have $\{u+[h\delta u_1 \oplus (1-h)\delta u_2]\} \in C$. The operative sum of two variations $\delta u_1(t) = a_k$ and $\delta u_2(t) = a_k''$, $t_k' \le t < t_k' + \varepsilon\Delta_k'$, $t_k' \le t < t_k' + \varepsilon\Delta_k''$, is defined by

$$\delta u_1(t) \oplus \delta u_2(t) = a_k', \qquad t_k' \le t < t_k' + \varepsilon\Delta_k',$$

$$\delta u_1(t) \oplus \delta u_2(t) = a_k'', \qquad t_k' + \varepsilon\Delta_k' \le t < t_k' + \varepsilon(\Delta_k'+\Delta_k'').$$

An admissible control A is called inferior to the admissible control B if the relation

$$y_i(t_2)\Big|_A \geq y_i(t_2)\Big|_B,$$

is satisfied, in which strict inequality holds for at least one of the indices i.

An admissible control is called unimprovable if it is not inferior to the rest of the admissible controls. An unimprovable admissible control defines an unimprovable system.

Theorem. Let there be given the fixed points $x(t_1)$ and $x(t_2)$ and let X be an unbounded space for x. In order that the admissible control $\hat{u}(t)$ and the trajectory corresponding to it $\hat{x}(t)$ be unimprovable, it is necessary that there exist a vector function $\hat{\Psi}(\hat{\Psi}_1,\ldots,\hat{\Psi}_n)$ and a function $H(\hat{\Psi},\hat{x},\hat{u},t)$ satisfying the following relation

$$H(\hat{\Psi},\hat{x},\hat{u},\hat{t}) \equiv \sum_{n=1}^{n} \hat{\Psi}_i(t) f_i(\hat{x},\hat{u},t) - \sum_{k=1}^{N} c_k g_k(\hat{x},\hat{u},t),$$

$$\partial_u \int_{t_1}^{t_2} H(\hat{\Psi},\hat{x},t)\,dt \leq 0, \tag{1.15}$$

$$\frac{d\hat{\Psi}_i(t)}{dt} = -\frac{\partial H}{\partial \hat{x}_i} \quad (i=1,\ldots,n),$$

$$c_k \geq 0 \quad (k=1,\ldots,N).$$

In the above expression (1.15) equality cannot hold for all values of k; by ∂_u is meant the first variation of the above integral corresponding to an infinitesimal variation of u(t) for fixed $\hat{\Psi}$, $\hat{x}$, t.

The formulated theorem bears the name the "general theorem of optimal control." It is proved in (54). We note the following: in (54), Chang obtained mathematical results formulated in the form of necessary conditions for unimprovable controls. However, the general theorem formulated above and the subsequent theorems from (54) cannot be successfully applied for the solution of concrete control problems with multiple criteria since the question of determination of the coefficients $c_k (k=1,\ldots,N)$ in (1.15) remains open.

The determination of unimprovable solutions is also encountered in statistical optimization problems. As was already noted, in (62) necessary and sufficient conditions for the existence of an unimprovable solution in mathematical programming problems is shown, but the author does not show the application of his results in practical problems.

Recently, unimprovable solutions have been called Pareto-optimal solutions.

6. The Work of A. Salama and V. Gourishankar

In (89) systems described by the equations

$$\dot{x} = f(x(t), u(t)), \tag{1.16}$$

are studied. The components of $f(x,u)$ are assumed to be continuous in u and smooth in x. The domain (1.2) is given by the conditions

$$\theta_i(u_1,\ldots,u_m) \leq \alpha_j \quad (j=1,\ldots,m), \tag{1.17}$$

in which the constants $\alpha_j > 0$. In this case, the set of admissible controls is the set of piecewise-continuous functions satisfying (1.17). It is assumed that the initial state of the system is fixed, i.e. (i): $x(0) = \text{const}$, while the terminal state (f) is given by the hyperplane $\Psi(x) = 0$.

The set of functionals to be optimized is represented in the form of Bolza

$$I_k = g_k(x(T)) + \int_0^T \ell_k(x(t),u(t))dt \leq \beta_k \qquad (k=1,\ldots,N), \qquad (1.18)$$

where $\beta_1,\ldots,\beta_N$ are given numbers. Here $g_1,\ldots,g_N$ and $\ell_1,\ldots,\ell_N$ are real, nonnegative functions, where $\ell_1,\ldots,\ell_N$ are continuously differentiable, while $g_1,\ldots,g_N$ are twice-continuously differentiable in x.

The authors introduce the function $\Phi(z) \in C^2$, which is defined in the following way: let $z\{z_1,\ldots,z_N\}$ be a vector defined by the equations

$$\dot{z}_k = \ell_k(x(t),u(t)) + \left(\frac{\partial g_k(x(t))}{\partial x}\right)' \cdot f(x(t),u(t)),$$

with the initial conditions

$$z_k(0) = g_k(x(0)).$$

Here the prime sign ' denotes the matrix transpose operation (the expression $\left(\frac{\partial g_k}{\partial x}\right)'$ represents a column matrix). Some system of nominal values z_{kd} $(k=1,\ldots,n)$ is introduced, as is a system of penalty functions $p_k = p_k(z_k - z_{kd})$. Next the function $\Phi(z)$ is expressed in the form

$$\Phi(z) = \sum_{k=1}^{N} p_k(z_k - z_{kd})$$

and it is proposed to minimze it with respect to u.

The essentially new element introduced in the given problem statement is the idea of approximate optimization of a vector functional. The measure of approximation is defined by the choice of the function $\Phi(z)$.

The new problem statement is also formulated in the works (142, 143), published at almost the same time as the work (89). With respect to the approach in (89), we make several critical remarks. First of all, the numbers $\beta_1, \ldots, \beta_N$ introduced in (1.18) do not play an important role in (89) and are not used in its subsequent development. They are unnecessary elements in the problem because for optimization of K_k under the constraint (1.18), they do not enter at any place in (89).

Secondly, in subsequent treatments of the original problem the authors of (89) step back from the original formulation and consider optimization of the linear functional $\sum_k c_k z_k$ and the selection of the weighting coefficients $c_1, \ldots, c_N$ is such that this functional assumes its extremum at the same points of the space z as the original functional $\Phi(z)$.

Finally, it is important to note that the numbers z_{kd} are not specified here and this is an essential omission. These numbers cannot be given arbitrarily. They have a unique acceptable definition which is given in (142).

7. Solutions in Fuzzy Neighborhoods

A very interesting idea involving the determination of a solution in a fuzzy neighborhood is described in (100). By this, we mean a solution process when the system equations, the constraints and the objective function (or any combination of them) are described by fuzzy sets. Fuzzy sets are defined as sets the boundaries of which are given imprecisely. For example, a fuzzy constraint may be given by the rule: "the value of the objective function is basically no larger than a," where a is a given constant; or "x must be near x_0," where x_0 is also a given constant. The source of fuzziness in these expressions lies in the underlined words. In other words, a fuzzy set is some collection of objects where there is no precise mathematical boundary between belonging and not belonging to the collection.

The following is a more precise definition.

Definition. Let $X = \{x\}$ denote a set of independent elements $x_1, x_2, \ldots, x_n$. Then we will call the set of pairs

$$A = \{(x; \mu_A(x))\}, \qquad x \in X,$$

a fuzzy set A in X, where $\mu_A(x)$ is a function describing the degree of membership of the element x in A. The function $\mu_A(x)$, called the membership function, transforms the space X into the membership space M. When the membership space M consists only of the two points 0 and 1, the set A becomes determinate and the membership function $\mu_A(x)$ reduces to the usual characteristic function of a set. As an example, we use the collection of pairs

$$A = \{(3;0.6),(4;0.8);(5;1.0),(6;1.0)(7;0.8),(8;0.6)\},$$

where $X = \{0,1,2,\ldots\}$ is the set of nonnegative integers.

In the following definition, the authors give the meaning of a solution in a fuzzy neighborhood: let a fuzzy objective function G and a fuzzy constraint C be given in the space of alternatives X. Then G and C form a solution D, which is defined as the fuzzy set obtained as the intersection of G and C.

Symbolically, the solution is described in the form

$$D = G \cap C \tag{1.19}$$

with the associated membership function

$$\mu_D = \mu_G \wedge \mu_C. \tag{1.20}$$

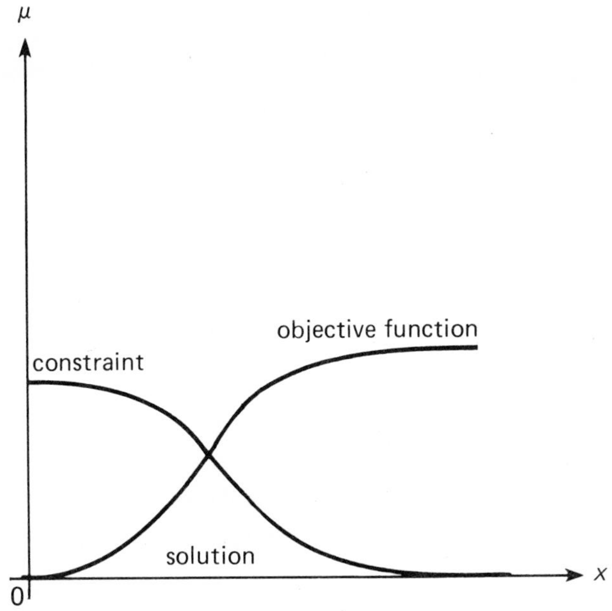

Fig. 3

The expression (1.19) means that D is the largest fuzzy set contained in both of the fuzzy sets G and C simultaneously, while the expression (1.20 defines the membership function of this largest set, where the symbol $\wedge$ is defined by the following rule:

$$\mu_D = \mu_{G \cap C}(x) = \begin{cases} \mu_G(x) & \text{for } \mu_G(x) \leq \mu_C(x) \\ \mu_C(x) & \text{for } \mu_C(x) > \mu_C(x), \quad x \in X. \end{cases}$$

The connection between G, C and D is depicted in Fig. 3.

The ideas just described extend to the more general case when there are several objective functions $G_1, G_2, \ldots, G_n$, say, and several constraints $C_1, C_2, \ldots, C_m$. In such a case, the authors define the fuzzy solution D as

$$D = G_1 \cap G_2 \cap \ldots \cap G_n \cap C_1 \cap C_2 \cap \ldots \cap C_m,$$

with the corresponding membership function

$$\mu_D = \mu_{G_1} \wedge \mu_{G_2} \wedge \cdots \wedge \mu_{G_n} \wedge \mu_{C_1} \wedge \mu_{C_2} \wedge \cdots \wedge \mu_{C_m}$$

It is not difficult to see that in the solution process the objective functions and the constraints play identical roles and the solution is represented as the product of the joining of the objective functions and the constraints.

We note that the presented method is the same as that in (59) where, actually, the set D turns out to be empty.

On the means of constructing the membership function of the fuzzy set A, the authors remain silent. In our opinion, for the solution of concrete problems it is very important to know a method for constructing such functions. We do not exclude the possibility of connecting the construction of such a function with the ideas for solving multicriteria optimization problems, which are presented in the succeeding chapters.

CHAPTER II
THE FORMULATION OF OPTIMIZATION PROBLEMS WITH VECTOR FUNCTIONALS

1. The General Mathematical Formulation of the Problem

We formulate a new statement of the problem of optimizing a vector functional. The idea of such a formulation was first presented in the author's works (142,143).

Let each element of some space Z define a system $S[Z]$, which is characterized by some vector performance index $I = I\{I_1,\ldots,I_k\}$. In such a situation each element of the space Z, defining the system $S[Z]$, will have associated with it a definite point with coordinates $I_1(Z),I_2(Z),\ldots,I_k(Z)$ in the space of performance indices $I(Z)$. So, for example, for the element $z \in Z$, the system $S_Z \in S[Z]$, will have the performance indices $I_1(z),\ldots,I_k(z)$ defining a point $I(z) \in I(Z)$.

We let $z_\alpha \in Z$ be an element of the space Z for which

$$I_\alpha(z_\alpha) = \sup_{z\in Z} I_\alpha(z),$$

with $\alpha = 1,\ldots,k$.

This means that for the element $z_\alpha \in Z$, only the performance index $I_\alpha(z)$ assumes its optimal value.

We consider the euclidean space $I(Z)$ with coordinates $I_1(Z),\ldots,I_k(Z)$. Let the values $I_\alpha(z_\alpha)$ $(\alpha=1,\ldots,k)$ be given. It is always possible to say that there exists an element $z^* \in Z$, forming a system $S_{z*} \in S[Z]$ which, in the space $I(Z)$, the values of the functionals $I_1(z^*),\ldots,I_k(z^*)$ will be the best (in some agreed upon sense) approximation to the true optimal values of these functionals $I_1(z_1), I_2(z_2), \ldots, I_k(z_k)$. As a criterion for such an approximation, it is possible to take any positive-definite function of the variables $I_1(Z), I_2(Z), \ldots, I_k(Z)$. We let $R(Z)$ denote such a function which forms one or another metric in the space $I(Z)$.

Definition. An element $z^* \varepsilon Z$ optimizes the system S in the $R(Z)$-sense if

$$R(z^*) = \inf_{z \varepsilon Z} R(z). \tag{2.1}$$

Consequently, the problem of optimizing the system S with a vector criterion may be formulated as: given $S[Z]$, $I(Z)$, $R(Z)$, find $z^* \varepsilon Z$.

As noted in the last chapter, the vector optimization problem was considered in much the same spirit in (89); however, it is important to emphasize the essential difference between the problem statement given above and that in (89). Recently, these ideas have been developed in the works (91-99).

2. A Statement of Optimal Trajectory Programming Problems with a Vector-Valued Criterion

We present a statement of the problem in the same form as it was given in (142).

Let the system dynamics be given by the differential equations

$$\dot{x} = f(x(t), u(t), t), \tag{2.2}$$

which are defined in some region $N(x(t), U(t)) \geq 0$ of state space $x(x_1, \ldots, x_n)$ and control space $u(u_1, \ldots, u_m)$, $t \varepsilon \tau$, $\tau = [t_0, T]$.

The right side of the expression (2.2) is a vector with coordinates $f_1(x,u,t), f_2(x,u,t), \ldots, f_n(x,u,t)$, which are assumed to be continuously differentiable functions in the variables x, u and t.

Let some admissible class of controls U be given where $u \in U$ assumes values in the region $N \geq 0$. Also, let there be given a vector functional

$$I(u) = \Phi(x(t),u(t),t) \tag{2.3}$$

with components

$$I_\alpha(u) = \Phi_\alpha(x(t),u(t),t) \quad (\alpha=1,\ldots,k). \tag{2.4}$$

It is assumed that the functions $\Phi_1, \Phi_2, \ldots, \Phi_n$ have continuous first partial derivatives with respect to all arguments in the region $N \geq 0$ for $t \in \tau$.

Let the boundary values for the vector x(t) be

$$x(t_0) = x_0, \qquad x(T) = x_T, \tag{2.5}$$

where T is free. Symbolically, we denote this as $(i,f) = 0$.

Using known procedures, for each $\alpha = 1,\ldots,k$ we determine the optimal control

$$u^{(\alpha)} = u^{(\alpha)}(t,x_0,x_T) \quad (\alpha=1,\ldots,k) \tag{2.6}$$

for each scalar functional $I_\alpha(u)$.

Here the vector $u^{(\alpha)}$ has components $u_1^{(\alpha)},\ldots,u_m^{(\alpha)}$ and is the optimal control vector under which the scalar functional $I_\alpha(u) = \Phi_\alpha(x(t),u(t),t)$ assumes its optimal value along the trajectory of the system (2.2) which passes between the boundary points (2.5). Naturally, there are different periods of time $T_1,\ldots,T_k$ associated with the control vectors $u^{(\alpha)}$ $(\alpha=1,\ldots,k)$. Moreover, when optimizing each scalar functional, additional conditions may be taken into account such as construction of the region $N \geq 0$, specific boundary conditions and also the class $V^{(\alpha)}$ of admissible controls.

We compute the value of the vector

$$I^*(u)\{I_1(u^{(1)}),I_2(u^{(2)}),\ldots,I_k(u^{(k)})\}. \tag{2.7}$$

Its components are known numbers.

We consider the euclidean norm

$$R(u) = ||I(u)-I^*(u)||^2 = \sum_{\alpha=1}^{k}\left[I_\alpha(u)-I_\alpha(u^{(\alpha)})\right]^2 \tag{2.8}$$

where the vector $I(u) - I^*(u)$ is defined for all admissible controls $u \in V$. Obviously, the set of admissible controls V must be the maximal restriction of all the admissible $V^{(1)},\ldots,V^{(k)}$.

Definition. We will say that the control $u^0(t,x_0,x_T)$ optimizes the vector functional (2.3) if the inequality

$$R(u^0) \leq R(u) \tag{2.9}$$

is satisfied for any admissible control u. We call such a control optimal relative to the vector functional.

Problem. Given the equation (2.2), the boundary conditions (2.5), the vector functional (2.3) and the class V of admissible controls, determine a control $u^0(t,x_0,x_T)$ optimizing the vector functional.

In the language of the preceding section, here the control vector u(t) plays the role of z and determines the motion of the system (2.2). As a criterion of approximation R(Z) we choose the expression (2.8), while the meaning of the equality (2.1) is embodied in the inequality (2.9).

Remark 1. It is assumed that in the original problem all the variables and functionals are reduced to dimensionless form. If not, then instead of minimizing the expression (2.8) we must consider minimizing the sum of squares of the relative deviation of the functionals (2.4) from their nominal optimal values (2.7), i.e. we minimize the function

$$R(u) = \left|\left|\frac{I(u)-I^*(u)}{I^*(u)}\right|\right|^2 = \sum_{\alpha=1}^{k}\left[\frac{I_\alpha(u)-I(u^{(\alpha)})}{I_\alpha(u^{(\alpha)})}\right]^2 \qquad (2.10)$$

A generalized form of the approximation function R(S) is to take the norm

$$R_L(u) = \left(\sum_{\alpha=1}^{k}\left[I_\alpha(u)-I_\alpha(u^{(\alpha)})\right]^L\right)^{\frac{1}{L}}, \quad L \geq 1, \qquad (2.11)$$

which, for L = 1, reduces to a linear combination of the components of the vector $I(u) - I^*(u)$, for L = 2 it coincides with the euclidean norm $||I(u)-I^*(u)||$ and for $L = \infty$ the supremum norm

$$R_\infty(u) = \max_\alpha \{I_\alpha(u)-I_\alpha(u^{(\alpha)})\,|\,\alpha=1,\ldots,k\}. \qquad (2.12)$$

Remark 2. It is natural to let T be either free or fixed. In particular, if the numbers $T_1,\ldots,T_k$ do not differ too greatly from each other, then the time T must also not deviate too greatly from these values. Moreover, it makes sense to consider the case when T is fixed and equals, for instance, $T = \frac{1}{k}\sum_{\alpha=1}^{k} T_\alpha$ or $T \leq \max\{T_\alpha\,|\,\alpha=1,\ldots,k\}$.

A geometrical interpretation of the problem. We consider the euclidean space for the vector I(u). The sum (2.8) is the square of the distance from the point corresponding to the control u to the point having coordinates $I_1(u^{(1)}), I_2(u^{(2)}), \ldots, I_k(u^{(k)})$. We call this latter point the ideal (or utopian) point. The problem consists in choosing a control u which minimizes this distance over a time T, which does not deviate too much from the numbers $T_1, \ldots, T_k$.

A physical interpretation of the problem. If the vector $u^0(t, x_0, x_T)$ minimizes (2.8), then it will be the vector under which the functionals $I_\alpha(u^0)$ ($\alpha=1,\ldots,k$) assume values as near as possible to the numbers $I_1(u^{(1)}), I_2(u^{(2)}), \ldots, I_k(u^{(k)})$.

The sense of this approximation is as follows: we assume that we have chosen some scalar functional I_α, determined the desired class of admissible controls $V^{(\alpha)}$ for it and found the optimal control $u^{(\alpha)}$. The number $I_\alpha(u^{(\alpha)})$ we take as the index of quality of the system, which may be achieved in an ideal case. The ideal case is that in which the total control resources are used for achieving the optimal value for only one of the criteria. Moreover, in this ideal case it is possible, in general, to follow two different directions characterized by the choice of the class of admissible controls and also by the choice of boundary conditions for optimization of each scalar functional taken individually. In actuality, we may not ignore the other performance indices and often, in order to achieve an improvement in the system relative to the set of all these indices, we must operate in a more restricted class of admissible controls.

For the chosen control u^0, there takes place some deterioration of each performance index $I_\alpha(u)$ if taken separately (in comparison with that value obtained if we would optimize only with respect to that particular criterion); however, this deterioration is spread throughout the whole set of criteria $I_\alpha(u)$ and is as small as possible.

Remark 3. The usual approach to the problem of optimizing k functionals is to minimize the sum $\sum_{\alpha} \lambda_{\alpha} I_{\alpha}(u)$. Here it is assumed that the weighting coefficients are given exactly, i.e. the weights of the functionals $I_{\alpha}(u)$ $(\alpha=1,\ldots,k)$ are known exactly. However, this is possible only in a few cases.

Let us now assume that the statement of the vector optimization problem is freed from the constraint of knowing the weights of the functionals. We seek a control which uniformly approximates the values of the functionals $I_{\alpha}(u^0)$ to their optimal values $I_{\alpha}(u^{(\alpha)})$. There are now opportunities for other forms of approximation.

Remark 4. Here we have formulated the problem of programming optimal trajectories. Usually, it is also possible to formulate the problem of analytic construction of a regulator $u^0(x,t)$, which optimizes the vector functional.

Again it is relevant to emphasize the essential difference of the given problem statement from that which was given in section 6 of Chapter I.

1. The choice of the numbers $I_1(u^{(1)}), I_2(u^{(2)}), \ldots, I_k(u^{(k)})$ is stipulated in a unique fashion. In the space of the vector $I\{I_1, I_2, \ldots, I_k\}$ this means the following (see Fig. 4 for a simple illustration of the case k = 2). Let the point I* have coordinates

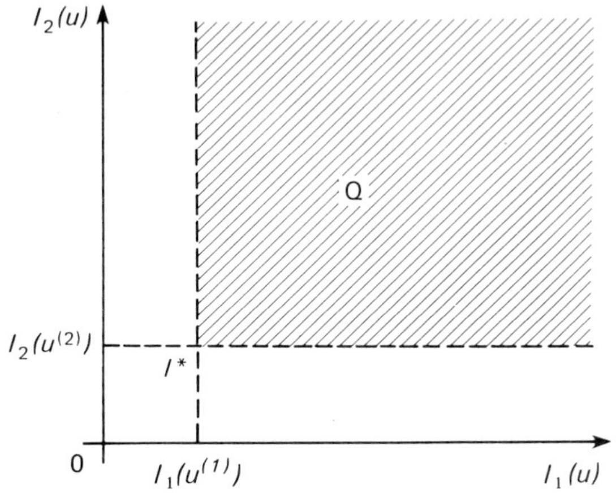

Fig. 4

$I_1(u^{(1)}), I_2(u^{(2)}), \ldots, I_k(u^{(k)})$. Then it is clear that optimization of the vector functionals (2.3) is possible only in the shaded positive octant Q. But this means that as a measure of approximation of the vector I(u) to the point I*, we may choose not only the euclidean norm (2.8) or (2.10), but also (2.11) for any $L \geq 1$ and the linear function

$$R(u) = \sum_{\alpha=1}^{k} c_\alpha \left[I_\alpha(u) - I_\alpha(u^{(\alpha)}) \right] \tag{2.13}$$

for constants $c_\alpha > 0$ given in advance. We can also use any continuous positive function R(u) defined on Q.

2. Assignment of the boundary values (2.5) may be carried out in a different manner for each optimization taking the functionals $I_\alpha(u)$ separately, but then for optimization of the function R(u) these conditions must reflect a basic and not auxiliary aspect of the problem.

3. In each particular optimization problem, the set of admissible controls is particular to that problem. Since the statement of the vector optimization problem is developed for the set of admissible controls V, which is itself restricted by $V^{(1)}, V^{(2)}, \ldots, V^{(k)}$, the position of the point I* depends upon the structures of the sets $V^{(1)}, V^{(2)}, \ldots, V^{(k)}$. The value $\min_{u \in V} R(u)$ will be minimal in the case when the point I* is also defined for $u \in V$.

3. Flight at a Prescribed Altitude

We illustrate the given problem statement by an example from the book (10). The equations of flight of an aircraft at a given altitude under reasonable assumptions may be expressed in the form (1.11).

We introduce the boundary conditions:

$$\begin{aligned}
&(\mathrm{i}): \quad x_i = 0, \qquad v_i = v_0, \qquad m_i = m_0, \\
&(\mathrm{f}): \quad x_f = x_T, \qquad v_f = v_T, \qquad m_f = m_T,
\end{aligned} \tag{2.14}$$

Here v_0, m_0, v_T, m_T are given numbers since the values x_T and T are free. As a functional we take

$$\Delta G = -x_T, \tag{2.15}$$

the minimum of which corresponds to maximizing the distance x_T.

Now among the admissible x, v, m, β we find that curve which minimizes the functional (2.15).

According to the maximum principle (6), we form the function

$$H = v\Psi_1 - \frac{Q}{m}\Psi_2 + \left(\frac{c}{m}\Psi_2 - \Psi_3\right)\beta \tag{2.16}$$

and write the equation for the co-state vector $\Psi\{\Psi_1, \Psi_2, \Psi_3\}$

$$\begin{aligned}
\dot{\Psi}_1 &= 0, \\
\dot{\Psi}_2 &= -\Psi_1 + \frac{Q_v}{m}\Psi_2, \\
\dot{\Psi}_3 &= \frac{1}{m}\left[Q_m + \frac{c\beta - Q}{m}\right]\Psi_2 .
\end{aligned} \tag{2.17}$$

Here

$$Q_v = \frac{\partial Q}{\partial v}, \qquad Q_m = \frac{\partial Q}{\partial m} .$$

A first integral and the transversality conditions are written in the form

$$v\Psi_1 - \frac{Q}{m}\Psi_2 + \left(\frac{c}{m}\Psi_2 - \Psi_3\right)\beta = C, \tag{2.18}$$

$$\left[(\Psi_1 - 1)\delta x - C\delta t + \Psi_2 \delta v + \Psi_3 \delta m\right]_i^f = 0. \tag{2.19}$$

respectively.

By virtue of the boundary conditions (2.14), condition (2.19) gives

$$\Psi_{1T} = 1, \quad C = 0. \tag{2.20}$$

For the optimal control, the maximum principle defines the solution

$$\beta = \bar{\beta} \text{ for } \frac{c}{m}\Psi_2 - \Psi_3 > 0,$$
$$\beta = 0 \text{ for } \frac{c}{m}\Psi_2 - \Psi_3 < 0. \tag{2.21}$$

The singular solution $0 \leq \beta \leq \bar{\beta}$ appears on an interval $[t_1, t_2]$ where we have the two integrals

$$v\Psi_1 - \frac{Q}{m}\Psi_2 = 0,$$
$$\frac{c}{m}\Psi_2 - \Psi_3 = 0. \tag{2.22}$$

By Poisson's theorem (c.f. (18), section 18), we obtain a third integral

$$cm\Psi_1 + (mQ_m - cQ_v - Q)\Psi_2 = 0. \tag{2.23}$$

Determination of the three relations (2.22), (2.23) requires

$$\begin{bmatrix} v & -\frac{Q}{m} & 0 \\ 0 & \frac{c}{m} & -1 \\ cm & mQ_m - cQ_v - Q & 0 \end{bmatrix} = 0. \tag{2.24}$$

This relation defines the manifold

$$S_1(m,v) = cQ + v(mQ_m - cQ_v - Q) = 0. \tag{2.25}$$

In the particular case of a parabolic polar, where

$$Q = Av^2 + B\frac{m^2}{v^2} \tag{2.26}$$

with A, B given positive constants, we will have

$$S_1(m,v) = Bm^2(v+3c) - Av^4(v+c) = 0. \tag{2.27}$$

It is not difficult to establish that here Poisson's theorem does not give any other linearly independent integrals. According to (18) (cf. §18), the form of the singular control is written as

$$\beta_1 = \frac{QS_{1v}}{cS_{1v} - mS_{1m}}, \quad \left(S_{1v} = \frac{\partial S_1}{\partial v}, \; S_{1m} = \frac{\partial S_1}{\partial m}\right), \tag{2.28}$$

which, in the case of the parabolic polar (2.26), will have the form

$$\beta_1 = \frac{\left[Av^4(3v+2c) + Bm^2(v+6c)\right]Q}{Acv^4(3v+2c) + Bm^2v(2v+7c) + 6Bc^2m^2}. \tag{2.29}$$

Thus, the optimal control of the thrust is represented, generally speaking, as a piecewise-continous function consisting of the three arcs (2.21), (2.28). As was already noted in section 3 of the preceeding chapter, the optimal combination of these arcs may be the curve impuf depicted in Fig. 5.

From the construction, it follows that in any case the control (2.28) satisfies the condition $0 \leq \beta_1 \leq \bar{\beta}$. Actually, since the manifold (2.25) passes through the origin, $\frac{dm}{dv} > 0$ and, consequently, for the control (2.28) $c\beta < Q(m,v)$ while at the time when $\beta = \bar{\beta}$ we have $c\beta > Q(m,v)$.

Now we define a program thrust which will be optimal in the minimum time sense, i.e. for the system performance index we take the functional

$$I_2 = \int_0^T dt = T. \tag{2.30}$$

We describe the problem in new variables. We let x be the independent variable. Then Eq. (1.11) takes the form

$$\begin{aligned} \frac{dt}{dx} &= \frac{1}{v}, \\ \frac{dv}{dx} &= \frac{c\beta - Q(m,v)}{mv}, \\ \frac{dm}{dx} &= -\frac{\beta}{v}. \end{aligned} \tag{2.31}$$

The functional (2.30) is written in the Mayer form as

$$G = t, \quad \Delta G = t_f = T_{min}. \tag{2.32}$$

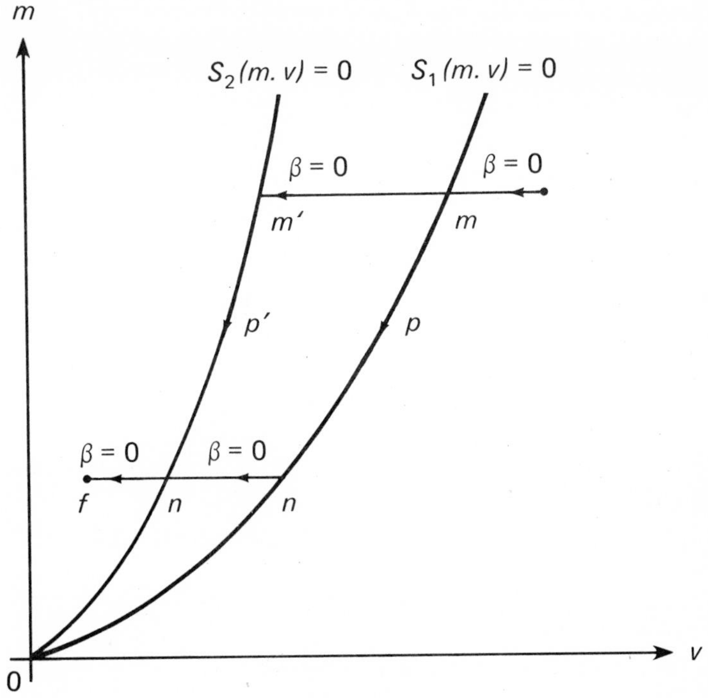

Fig. 5

We now apply the maximum principle. We have the function

$$H = \frac{\Psi_1}{v} - \frac{Q}{mv}\Psi_2 + \left(\frac{c}{mv}\Psi_2 - \frac{\Psi_3}{v}\right)\beta, \tag{2.33}$$

the auxiliary equations

$$\begin{aligned} \frac{d\Psi_1}{dx} &= 0, \\ \frac{d\Psi_2}{dx} &= \frac{\Psi_1}{v^2} + \frac{vQ_v - Q}{mv^2}\Psi_2 + \left(\frac{c}{mv^2}\Psi_2 - \frac{\Psi_3}{v^2}\right)\beta \\ \frac{d\Psi_3}{dx} &= \frac{mQ_m - Q}{m^2 v}\Psi_2 + \frac{c\beta}{m^2 v}\Psi_2 \end{aligned} \tag{2.34}$$

and the form of the optimal thrust as

$$\beta = 0 \text{ for } \frac{c}{mv}\Psi_2 - \frac{1}{v}\Psi_3 < 0,$$
$$\beta = \bar{\beta} \text{ for } \frac{c}{mv}\Psi_2 - \frac{1}{v}\Psi_3 > 0. \tag{2.35}$$

The transversality conditions determine $\Psi_{1T} = -1$ and $C = 0$.

We construct the singular control. Since x is free, we have the first integral

$$P + M\beta = 0,$$

where P and M are

$$P = \frac{\Psi_1}{v} - \frac{Q}{mv}\Psi_2, \quad M = \frac{c}{mv}\Psi_2 - \frac{\Psi_3}{v}. \tag{2.36}$$

On the interval $[t_1, t_2]$, the singular solution satisfies the two integrals

$$P = \frac{\Psi_1}{v} - \frac{Q}{mv}\Psi_2 = 0,$$
$$M = \frac{c}{mv}\Psi_2 - \frac{\Psi_3}{v} = 0. \tag{2.37}$$

The Poisson bracket gives the new first integral

$$\left(-\frac{\Psi_1}{v^2} - \frac{vQ_v - v}{mv^2}\Psi_2\right)\frac{c}{mv} + \frac{Q}{mv}\left(-\frac{c}{mv^2}\Psi_2 + \frac{\Psi_3}{v^2}\right) + \frac{1}{v}\frac{mQ_m - Q}{m^2 v}\Psi_2 = 0, \tag{2.38}$$

or, what is equivalent

$$\left[-mc\Psi_1 + v(mQ_m - cQ_v - Q)\Psi_2 + mQ\Psi_3\right] = 0. \tag{2.39}$$

The integral manifold of the singular solutions must satisfy the equation

$$\begin{vmatrix} \frac{1}{v} & -\frac{Q}{mv} & 0 \\ 0 & -\frac{c}{mv} & -\frac{1}{v} \\ -cm & v(mQ_m - cQ_v - Q) & mQ \end{vmatrix} = 0. \tag{2.40}$$

Thus, as follows from (2.43), we have the manifold of singular solutions

$$S_2(m,v) = mQ_m - cQ_v - Q = 0, \tag{2.41}$$

which, in the case of the parabolic polar (2.26), is represented by the expression

$$S_2(m,v) = Bm^2 - Av^4 = 0. \tag{2.42}$$

Analagously, by the formula (2.28), the form of the singular control is written as

$$\beta_2 = \frac{QS_{2v}}{cS_{2v} - mS_{2m}}, \quad \left(S_{2v} = \frac{\partial S_2}{\partial v}, \; S_{2m} = \frac{\partial S_2}{\partial m} \right). \tag{2.43}$$

In the case of the parabolic polar (2.26), the singular control may be written as

$$\beta_2 = \frac{2Av^3Q}{2Acv^3 - Bm^2}. \tag{2.44}$$

Turning to graphics, the manifold $S_2(m,v) = 0$ will be located to the left of the manifold $S_1(m,v) = 0$, as is shown in Fig. 5. In order to convince ourselves of this, from formulas (2.25) and (2.41) we determine Q. In the case $S_1(m,v) = 0$, we have

$$Q = Q_1(m,v) = \frac{v}{v - c}(mQ_m - cQ_v), \tag{2.45}$$

while in the case of the manifold $S_2(m,v) = 0$,

$$Q = Q_2(m,v) = mQ_m - cQ_v. \tag{2.46}$$

Since the strength of the aerodynamic drag $Q(m,v)$ is positive and $c > 0$, the comparison of (2.45) with (2.46) gives

$$Q_1(m,v) > Q_2(m,v). \tag{2.47}$$

Since the mass only decreases, for a fixed mass, equality of Q_1 and Q_2 can only be achieved under the condition $v_1 > v_2$. The validity of this is shown by the arrangement of the curves $S_1 = 0$ and $S_2 = 0$ in Fig. 5.

Such an ordering becomes evident for determination of Q in the form of the parabolic polar (2.26). Describing the manifold (2.27) in the form

$$m^2 = \frac{A}{B} v^4 \frac{v + c}{v + 3c},$$

and the manifold (2.42) in the form

$$m^2 = \frac{A}{B} v^4,$$

it is evident that for a fixed value of the velocity we will always have $m_1 < m_2$.

In the first problem for the functional (2.15), the optimal trajectory, as was noted, corresponds to the curve impnf, while in the second problem for the functional (2.32), the optimal curve is im'p'n'f (Fig. 5).

Now we consider the motion which is optimal, say, for the criterion

$$G = \frac{1}{2}\left(\frac{x}{x_0} - 1\right)^2 + \frac{1}{2}\left(\frac{t}{t_0} - 1\right)^2, \tag{2.47'}$$

in which x^0 is the maximal achieved distance in the first optimization problem, while t^0 is the minimal time in the second problem.

The optimal control will be composed of $\beta = 0$, $\beta = \bar{\beta}$ and singular arcs.

In fact, the Pontryagin function has the same form as in (2.16); the auxiliary equations also coincide with (2.17) and the first integral remains in the form (2.18). This insures that the form of the optimal control is as defined in (2.21); however, for determination of the vector Ψ we use the following transversality condition

$$\left[\left(\frac{x-x^0}{(x^0)^2} + \Psi_1\right)\delta x + \left(\frac{t-t_0}{(t^0)^2} - C\right)\delta t + \Psi_2\delta v + \Psi_3\delta m\right]_i^f = 0. \tag{2.48}$$

According to the boundary conditions (2.14), the condition (2.48) splits into the relations

$$\left(\frac{x-x^0}{(x^0)^2} + \Psi_1\right)_{t=T} = 0, \tag{2.49}$$

$$\left(\frac{t-t^0}{(t^0)^2} - C\right)_{t=T} = 0. \tag{2.50}$$

Now the first integral is written as

$$P - M\beta = \frac{T - t^0}{(t^0)^2}, \tag{2.51}$$

in which T is the duration of the process, while P and M have the forms

$$P = v\Psi_1 - \frac{Q}{m}\Psi_2, \qquad M = \frac{c}{m}\Psi_2 - \Psi_3. \tag{2.52}$$

On the interval $[t_1, t_2]$, the singular solution will satisfy the following integrals

$$P = v\Psi_1 - \frac{Q}{m}\Psi_2 = \frac{T - t^0}{(t^0)^2} = \text{const} = \bar{C} \geq 0, \tag{2.53}$$

$$M = \frac{c}{m}\Psi_2 - \Psi_3 = 0. \tag{2.54}$$

We calculate the first Poisson bracket

$$R = (P,M) = \frac{cm\Psi_1 + N\Psi_2}{m^2} = 0, \tag{2.55}$$

in which

$$N(m,v) = Q_m m - cQ_v - Q = S_2(m,v). \tag{2.56}$$

Now we calculate the second Poisson bracket

$$(R, P+M\beta) = (R,P) + (R,M)\beta = 0, \tag{2.57}$$

which, as a result, defines the new integral

$$-m(c\beta+N)\Psi_1 + \left[(cN_v - mN_m)\beta + NQ_v - N_v Q\right]\Psi_2 = 0. \tag{2.58}$$

From the integrals (2.54), (2.55) and (2.58), we determine the manifold of singular solutions. We have the determinant

$$\begin{vmatrix} 0 & \frac{c}{m} & -1 \\ cm & N & 0 \\ -m(c\beta+N) & (cN_v - mN_m)\beta + NQ_v - QN_v & 0 \end{vmatrix} = 0, \tag{2.59}$$

which after expansion gives

$$S = -N^2 - c(NQ_v - QN_v) - c\left[N + (cN_v - mN_m)\right]\beta = 0. \tag{2.60}$$

From the last expression, we determine the singular control as

$$\beta = \frac{N^2 + c(NQ_v - QN_v)}{c(mN_m - cN_v - N)} . \tag{2.61}$$

Taking into consideration (2.56), the last expression may be written in the form

$$\beta = \frac{QS_{2v}}{S_2 + cS_{2v} - mS_{2m}} + \frac{S_2(Q - mQ_m)}{c(S_2 + cS_{2v} - mS_{2m})} . \tag{2.62}$$

We note that if in (2.52), $S_2 = 0$, then it is converted into (2.43), i.e. into the expression which was defined for the solution of the minimum time problem.

The relation

$$S_1 = cQ + vS_2,$$

obtained by comparing (2.25) and (2.41), gives the possibility to determine the dependencies

$$S_{2v} = \frac{1}{v}(S_{1v} + Q - mQ_m),$$

$$S_{2m} = \frac{1}{v}(S_{1m} - cQ_m),$$

substitution of which into (2.62) determines the singular control β in the following form:

$$\beta = \frac{QS_{1v}}{cS_{1v} - mS_{1m} + S_1} + \frac{S_1(Q - mQ_m)}{c(cS_{1v} - mS_{1m} + S_1)} . \tag{2.63}$$

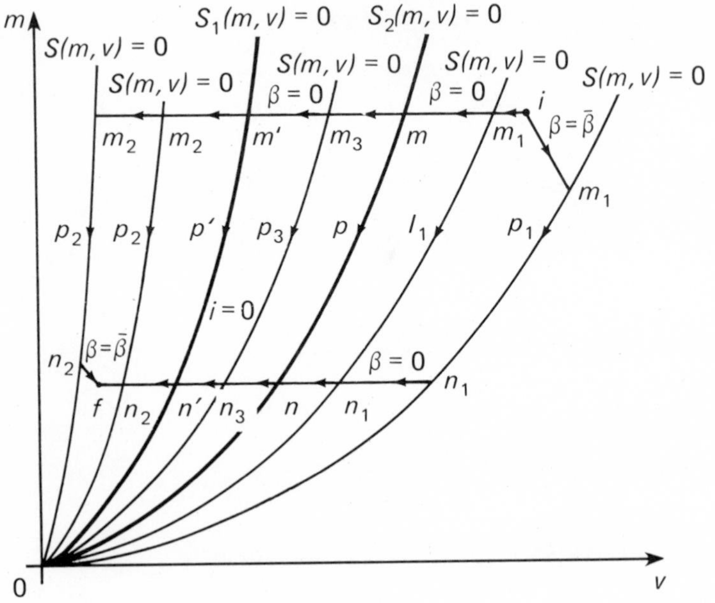

Fig. 6

If we suppose that $S_1 = 0$ in (2.63), then it is converted into (2.28), i.e. into the expression which was defined for the solution of the maximal distance problem.

The location of the curve (2.60) in the (m,v) plane depends on the form of the function Q(m,v), characterizing the aerodynamic drag. It may be situated to the right of the curve $S_1(m,v) = 0$ or to the left of the curve $S_2(m,v) = 0$, or between them. The motion of the point from i to f in this plane will be the curves $\underline{im_1p_1n_1f}$, $\underline{im_2p_2n_2f}$, $\underline{im_3p_3n_3f}$, respectively (Fig. 6).

We hope that this example has completely cleared up the expediency of the approach to the vector optimization problem in the spirit of section 1 of this chapter.

4. Analytic Construction of Optimal Regulators

The problem of analytic construction of optimal regulators (the ACOR problem) has the same statement as the problem of programming optimal trajectories. However, within it are specifications which demand its separate consideration.

First of all, we assume that we know the programmed control

$$x = x^*(t,x_0,x_T), \quad u = u^*(t,x_0,x_T), \tag{2.64}$$

which represents the exact solution of the vector equation (2.2) with boundary conditions(2.5).

The ACOR problem is posed relative to the perturbed dynamics described by the equation (18)

$$\dot{\eta} = B\eta + m\xi + \Theta(\eta,\xi,t) + \phi(t), \tag{2.65}$$

where η is the deviation of the vector x from the programmed value x^*, generated by means of the disturbance $\phi(t)$; ξ is the increment in the control vector used to bring the actual motion nearer to the programmed motion, i.e.

$$\eta_i(t) = x_i(t) - x_i^*(t), \quad \xi_j(t) = u_j(t) - u_j^*(t)$$

$$(i=1,\ldots,n;\ j=1,\ldots,m).$$

Here the matrices $B = \dfrac{\partial f(x,u,t)}{\partial x}$, $m = \dfrac{\partial f(x,u,t)}{\partial u}$ are known, time-varying functions defined on the motion (2.2) and calculated for the given program (2.64); by $\Theta(\eta,\xi,t)$ we denote a power series convergent for all η, ξ.

We impose a series of requirements on the perturbed motion given by Eq. (2.65). These requirements are based upon physical aspects of the controlled system (cf.(39)).

Let

$$I_\alpha(\xi) = \int_{t_0}^{T} \omega_\alpha(\eta,\xi)dt, \quad (\alpha=1,\ldots,k) \tag{2.66}$$

be functionals reflecting these requirements. This collection of functionals forms a vector functional.

The integrand functions $\omega_1(\eta,\xi),\ldots,\omega_k(\eta,\xi)$ are assumed to be positive-definite and continuously-differentiable in their arguments.

Let $||\eta_0|| \leq A$ be a neighborhood of the origin in which the perturbed motion can begin.

Let

$$\xi^{(\alpha)} = \xi^{(\alpha)}(\eta,t) \tag{2.67}$$

be a feedback law lying in the class of admissible controls V, yielding the optimum of the index I_α for the closed-loop system

$$\dot{\eta} = B\eta + m\xi^{(\alpha)} + \Theta(\eta,\xi^{(\alpha)},t) + \phi(t). \tag{2.68}$$

The law (2.67) may be synthesized for any $\alpha = 1,\ldots,k$, using known methods of solution of the ACOR problem for a single functional (18).

We note that the class of admissible controls V consists only of those functions $\xi(\eta,t)$ for which the system of equations (2.65) is asymptotically stable.

It is clear that for each fixed α, the law (2.67) may ignore all requirements contained in the other functionals of (2.66).

We assume that all the functions (2.67) are known and we compute the numerical value of all functionals $I_\alpha(\xi^{(\alpha)}) = I_{\alpha 0}$.

We form the euclidean norm in the space of vectors $I\{I_1,I_2,\ldots,I_k\}$:

$$R(\xi) = \sum_{\alpha=1}^{k} \left[I_\alpha(\xi) - I_\alpha(\xi^{(\alpha)}) \right]^2 = \sum_{\alpha=1}^{k} \left[I_\alpha(\xi) - I_{\alpha 0} \right]^2, \tag{2.69}$$

defined for all admissible controls $\xi \in V$.

Definition. We will say that the feedback law $\xi^0(\eta,t)$ optimizes the vector functional with components (2.66) if the inequality

$$R(\xi^0) \le R(\xi) \tag{2.70}$$

is satisfied for any admissible control $\xi \in V$. We will say that such a feedback law is optimal relative to the vector functional.

Problem. Given Eq. (2.65), the vector functional (2.66) and the class of admissible controls V, it is required to determine a feedback law $\xi^0(\eta,t)$ optimizing the vector functional in the above sense. This problem is studied in (143,148).

The mathematical formulation of the given problem is no different than the formulation of the problem of programming optimal trajectories considered in section 2 of this chapter. The distinction lies only in the given boundary conditions: in the ACOR problem all components of the state vector at the endpoints of the trajectory are assumed to be free and only in the special case of the infinite-interval problem must we impose the condition $\eta(\infty) = 0$.

However, there is a real difference in the form of the solutions to these two problems. In the problem of determining optimal programmed trajectories, the solution is sought in the form of a time-varying function of the boundary conditions, while in the ACOR problem the feedback law is sought in the form of a vector function of the state and time.

It is clear that all the remarks made relative to the choice of the function R(u) in section 2 of this chapter retain their meaning and value for the synthesis problem.

CHAPTER III
THE EXISTENCE OF SOLUTIONS IN OPTIMIZATION PROBLEMS WITH VECTOR-VALUED CRITERIA

1. Formulation of the Existence Problem

Let there be given a collection of smooth functions $G_\alpha(x)$ $(\alpha=1,\ldots,k)$. We assume that for each such function, taken individually, we have the statement of the problem of Mayer about programming the optimal trajectory for the controlled object described by the equations

$$\dot{x} = f(x(t),u(t)), \tag{3.1}$$

defined in the domain

$$N(x(t),u(t)) \geq 0 \tag{3.2}$$

for $t \in \tau$, $\tau = [0,T]$. We call each such problem a problem of scalar optimization.

We write the boundary conditions in symbolic form

$$(i,f)_\alpha = 0 \quad (\alpha=1,\ldots,k). \tag{3.4}$$

As noted in Chapter II, they may be either the same or different for different α. In particular, there may be the case of fixed origin (i) and free end (f) for free T, or the case where these endpoints are situated on several geometric manifolds. These boundary conditions are characteristic for problems of the state (i) to the state (f). Once again we note that such a problem is called a problem of programming trajectories. In other cases, it is characteristic to have a free initial condition (i) for fixed $t_i = 0$ and a fixed condition (f) for free T. This problem, as is well known, is called a regulator problem. These problems are of special interest for the case $T = \infty$.

We confine our consideration to the programming problem (149); extension to the regulator problem does not pose any essential difficulties.

How to pose the problem of existence?

We assume that for each $\alpha = 1,\ldots,k$, taken individually, there exists a solution to the programming problem

$$u^{(\alpha)} = u^{(\alpha)}(t,(i,f)_\alpha),$$

$$x^{(\alpha)} = x^{(\alpha)}(t,(i,f)_\alpha) \quad (\alpha=1,\ldots,k). \tag{3.4}$$

Let $T_1,T_2,\ldots,T_k \leq T$ be the time intervals for each individual problem of Mayer.

In the language of section 1 of Chapter I, the expressions (3.4) represent elements z_α of the space Z.

Let G_α^0 be the value of the functional G_α computed using the functions (3.4), i.e.

$$G_\alpha^0 = G_\alpha(x^{(\alpha)}) = I_\alpha(u^{(\alpha)}) \quad (\alpha=1,\ldots,k). \tag{3.5}$$

We note also that (3.5) is the same as $I_\alpha(z_\alpha)$ (cf. section 1, Chapter II).

We consider the vector $y\{y_1,\ldots,y_k\}$

$$y_\alpha = G_\alpha - G_\alpha^0 \quad (\alpha=1,\ldots,k). \tag{3.6}$$

It is obvious that the vector y is defined in the positive orthant of the space y. We consider the smooth function

$$R = R(y), \tag{3.7}$$

assuming only positive values for $y \neq 0$ and $R(0) = 0$.

As was already stated in the preceding chapter, the function $R(y)$ may be considered as a measure of approximation of the vector y to zero, if we seek the point y^* minimizing the function $R(y)$.

Below we give a concrete example of the function (3.7).

Thus, for example, if the numbers λ_α are positive then as a function $R(y)$ we may take

$$R_2(y) = \sum_{\alpha=1}^{k} \lambda_\alpha y_\alpha. \tag{3.8}$$

As another example, we may use the function

$$R_2(y) = \sum_{\alpha=1}^{k} \lambda_\alpha y_\alpha^2. \tag{3.9}$$

A more general form of (3.7) is given in (2.11).

We may use any smooth function (3.7) for the formation of the Mayer functional in the vector optimization problem. Clearly, it is also possible to choose a function

$$R(x,y) = R(y)\phi(x), \tag{3.10}$$

where $\phi(x)$ is any smooth, positive function of x.

Obviously, as a vector optimization functional we may choose a Lagrange functional of the form

$$\overline{R}(y) = \int_0^T R(y)\,dt \tag{3.11}$$

or

$$\overline{R}(x,y) = \int_0^T R(y)\phi(x)\,dt. \tag{3.12}$$

The basic question of the existence of a solution to the vector optimization problem is formulated as follows: let the function (3.7) be used as the degree of approximation of the vector G to the vector $G^0\{G_1^0,\ldots,G_k^0\}$ or, what is the same thing, the vector y to zero. It is required to establish whether or not there exists a solution to the problem of Mayer for each functional G_α, taken individually, where the Mayer problem is characterized by the relations

$$\begin{aligned} &\dot{x} = f(x(t),u(t)),\\ &N(x,u) \geq 0,\\ &(i,f) = 0,\\ &G(x) = R(y). \end{aligned} \tag{3.13}$$

It is assumed in all cases that the admissible controls belong to the class of piecewise-continuous functions taking on values from $N(x,u) \geq 0$.

We will attempt to answer the stated question in the case when the boundary conditions (3.3) are the same for all $\alpha = 1,\ldots,k$ and for N: $|u| \leq \bar{u}$.

2. The Maximum Principle for the Problem of Mayer with a Scalar Functional

Let the expression

$$H_\alpha(x,\Psi,u) = \sum_{i=1}^{n} f_i(x,u)\psi_i^{(\alpha)} \qquad (\alpha=1,\ldots,k) \tag{3.14}$$

define the Hamiltonian for fixed α for the scalar Mayer problem (18) with functional $G_\alpha(x)$.

Here, as already noted, the optimization problem with scalar performance index will, for brevity, be called the scalar optimization problem, while the multicriteria case is the vector optimization problem.

Here the auxiliary multipliers satisfy the equation

$$\dot{\Psi}_j^{(\alpha)} = -\sum_{i=1}^{n} \frac{\partial f_i}{\partial x_j} \Psi_i^{(\alpha)} \qquad (j=1,\ldots,n). \tag{3.15}$$

In addition, we have the transversality condition

$$\left[\sum_{\beta=1}^{n} \frac{\partial G_\alpha}{\partial x_\beta} \delta x_\beta - H_\alpha \delta t + \sum_{\beta=1}^{n} \Psi_\beta^{(\alpha)} \delta x_\beta\right]_i^f = 0. \tag{3.16}$$

For the programming problem, condition (3.16) gives the relations

$$H_\alpha(x^{(\alpha)}(T), \Psi^{(\alpha)}(T), u^{(\alpha)}(T)) = 0 \qquad (\alpha=1,\ldots,k), \tag{3.17}$$

$$\Psi_\beta^{(\alpha)}(T) = -\left(\frac{\partial G_\alpha}{\partial x_\beta}\right)_T \qquad \begin{pmatrix}\beta=1,\ldots,n\\ \alpha=1,\ldots,k\end{pmatrix}. \tag{3.18}$$

The maximum principle in each case, i.e for each α, determines a candidate optimal control

$$u^{(\alpha)} = u^{(\alpha)}(\Psi^{(\alpha)}, x^{(\alpha)}) \qquad (\alpha=1,\ldots,k). \tag{3.19}$$

Each $u^{(\alpha)}$ lies in the class of piecewise-continuous functions and assumes values in the region N.

As the solution (3.19) provides only $\max_u H$, it is essential to note that the form of the solution (3.19) is invariant relative to the form of the optimization functional G_α, i.e. the structure of the function $u^{(\alpha)}$ does not depend upon G_α. However, $\Psi^{(\alpha)}$ and $x^{(\alpha)}$ do depend upon G_α; consequently, $u^{(\alpha)}_{(x)}$ will be different for different $G_\alpha(x)$.

According to the statement of the problem, the collection of equations

$$\dot{x}_j^{(\alpha)} = f_j(x^{(\alpha)}, u^{(\alpha)}(\Psi^{(\alpha)}, x^{(\alpha)})), \tag{3.20}$$

$$\dot{\Psi}_j^{(\alpha)} = \sum_{i=1}^{n} \left(\frac{\partial f_i}{\partial x_j}\right)^{(\alpha)} \Psi_i^{(\alpha)} \qquad (j=1,\ldots,n) \tag{3.21}$$

are integrable for any boundary values $\Psi_\beta^{(\alpha)}(T)$. Actually, since Eqs. (3.21) are linear in Ψ, while the coefficients are continuous functions, the equations are integrable for any $\Psi_\beta^{(\alpha)}(0)$. Consequently, for any condition (3.18) for the vector Ψ at $t = T$, we can find a $\Psi_\beta^{(\alpha)}(0)$ so that the condition is satisfied.

Condition (3.17) determines the time T_α of duration of the solution (3.19).

Integration of the system of equations (3.20), (3.21), gives the optimal trajectory (3.4) corresponding to the interval $0 \leq t \leq T_\alpha$.

By assumption, such a solution exists for any $\alpha(\alpha=1,\ldots,k)$. It is important to note that the form of the functional G_α shows up only in the values of the vectors $\Psi^{(\alpha)}$ at the right end of the time interval, i.e. in the vectors $\Psi^{(\alpha)}(T)$. Clearly, this shows up in the form of the functions $\Psi^{(\alpha)}(t,\Psi^{(\alpha)}(T))$. However, whatever these functions may be, the controls $u^{(\alpha)}(t,(i,f)_\alpha)$ will be admissible and, consequently, the solution (3.4) will exist.

3. The Maximum Principle for Vector Optimization Problems

In the case of minimization of the functional (3.7), the problem of Mayer will be characterized by the relations (3.13).

For this problem, the Hamiltonian is given by

$$H = \sum_{i=1}^{n} f_i(x,u)\Psi_i, \tag{3.22}$$

where the Ψ_i are determined in form by the equations

$$\dot{\Psi}_j = -\sum_{i=1}^{n} \frac{\partial f_i}{\partial x_j} \Psi_i \qquad (j=1,\ldots,n). \tag{3.23}$$

The transversality condition is

$$\left[\sum_{\beta=1}^{n} \frac{\partial R}{\partial x_\beta} \delta x_\beta - H\delta t + \sum_{\beta=1}^{n} \Psi_\beta \delta x_\beta\right]_i^f = 0. \tag{3.24}$$

Hence, for the free endpoint T we have

$$H(x(T),\Psi(T),u(T)) = 0, \tag{3.25}$$

$$\Psi_\beta(T) = -\left(\frac{\partial R}{\partial x_\beta}\right)_T \qquad (\beta=1,\ldots,n). \tag{3.26}$$

The condition (3.25) determines the time T.

In order to answer the question of the existence of a solution

$$u^0 = u^0(t,(i,f)) \tag{3.27}$$

of the vector optimization problem, it suffices to compare:

1) the functions H and H_α,
2) the equations (3.23) and (3.15),
3) the conditions (3.26) and (3.18).

Since the form of the functions $u^{(\alpha)}$, (3.19), is invariant relative to the functional G_α and the functions $u^{(1)}, u^{(2)}, \ldots, u^{(k)}$ are assumed to exist, then the solution

$$u^0 = u^0(\Psi,x) \tag{3.28}$$

will also exist for the vector optimization problem. This means that in the space $u(x,\Psi)$, the functions H_α, H assume their maximum at the same point. In other words, since the functions H_α, H do not depend upon the form of the optimization functional, they take on their maximum on the same function $u(x,\Psi)$.

Since Eqs. (3.1), (3.23) coincide with Eqs. (3.20), (3.21) and the latter are integrable for the boundary conditions $(i,f)_\alpha$ and any values $\Psi^{(\alpha)}(0)$, the equations

$$\dot{x} = f(x,u^0(\Psi,x)),\qquad \dot{\Psi} = -\frac{\partial H}{\partial x}, \tag{3.29}$$

will be integrable, where for the function $u^0(t)$ we substitute the expression (3.28).

Remark. It is well known that the choice of boundary conditions is not arbitrary in any variational problem. This choice is determined by the physical requirements arising from the problem statement, and also from the form of the optimization functional. This situation is entirely transferable to the case of vector optimal control. As noted in Chapter II, the boundary conditions (3.3) in the scalar optimization problem, defining an ideal (utopian) point in the space of functionals $I_1,\ldots,I_k$, are chosen from the conditions of using all control resources for attaining the optimal values of each performance index of the system, taken individually. Thus, generally speaking, these conditions may be different. In the case of problem (3.13), the boundary condition $(i,f) = 0$ must be chosen so that the optimization function $R(y)$ makes sense. However, in this case the problem of the existence of a solution is not very complicated since the initial value of the co-state vector Ψ does not play a role in the form of the linear equations (3.15), (3.23) for the variables $\Psi_1,\ldots,\Psi_n$.

If we substitute the values of the integrals of Eq. (3.29) into (3.28), we obtain the solution of the vector optimization problem in the form (3.27). Thus, we can formulate the following result: if the vector functions $G(x)\{G_1(x), G_2(x), \ldots, G_k(x)\}$ are such that there exists an extremal solution of each scalar problem of Mayer taken separately, then for a differentiable function R there also exists an extremal solution of the vector problem of Mayer.

In other words, if the vector functional (2.3) in the problem of programming optimal trajectories is such that for each fixed component (2.4) the necessary conditions of optimality are satisfied (the maximum principle), then these conditions are satisfied if and only if the function (3.7) is differentiable in $x_1, x_2, \ldots, x_n$.

4. The Flight of a Pilotless Aircraft

The solution of the pilotless aircraft problem was given in (18). The aircraft's equations of motion have the form

$$\begin{aligned} \dot{x} &= u, \\ \dot{y} &= v, \\ \dot{u} &= \frac{c\beta}{m} \cos \omega, \\ \dot{v} &= \frac{c\beta}{m} \sin \omega - g, \\ \dot{m} &= -\beta. \end{aligned} \tag{3.30}$$

Here x is the range of the aircraft, y is the altitude, u is the horizontal component of the velocity, v is the vertical component of the velocity, m is the mass of the aircraft, together with its fuel, β is the rate of fuel consumption, ω is the angle of thrust relative to the horizontal, g is the acceleration of gravity and c is a given constant.

These equations are defined in a region $N \geq 0$ containing the constraints

$$0 \leq \beta \leq \bar{\beta}, \quad y \geq 0. \tag{3.31}$$

A problem in which β is not bounded is devoid of physical and mathematical sense.

Equations (3.30) are defined on the finite time interval

$$\tau = \left[0,T\right], \tag{3.32}$$

where T is unknown and cannot be given in advance.

We consider several different formulations of the problem.

Given the boundary conditions:

$$t_i = 0: \quad x_i = 0, \; y_i = 0, \; u_i = 0, \; v_i = 0, \; m_i = m_0, \tag{3.33}$$

i.e. the aircraft starts at the airport. The conditions at the right end are

$$t_f = T: \quad x_f = x_T, \; y_f = 0, \tag{3.34}$$

i.e. we consider landing the aircraft at a given point.

The quantities T, u_T, v_T, m_T remain free.

We let (a) denote the minimal fuel expenditure problem under conditions (3.30) - (3.34), i.e. for (a):

$$G = -m, \quad \Delta G = -(m_T - m_0). \tag{3.35}$$

We let (b) denote the minimal flight-time problem under the same conditions, i.e. for (b):

$$G = t, \quad \Delta G = T. \tag{3.36}$$

We let (c) be the problem of optimizing a vector functional with components (3.35), (3.36) for the same mathematical model (3.30)-(3.34), i.e. for (c):

$$G = R(m,t), \quad \Delta G = R(m_T,T) - R(m_0,0); \tag{3.37}$$

where the function $R(m,t)$ is a positive-definite, continuous function of its arguements.

We consider the solution of all three problems simultaneously.

We form the function H. In all three of the cases (a), (b), (c):

$$H = \Psi_1 u + \Psi_2 v - \Psi_4 g + k_\beta \beta. \tag{3.38}$$

Here, for the sake of compactness, we have introduced the notation

$$k_\beta = \frac{c}{m} k_\omega - \Psi_5, \tag{3.39}$$

$$k_\omega = \Psi_3 \cos \omega + \Psi_4 \sin \omega. \tag{3.40}$$

The co-state equations for the vector Ψ are written in the form

$$\dot{\Psi}_1 = 0,$$

$$\dot{\Psi}_2 = 0,$$

$$\dot{\Psi}_3 = - \Psi_1,$$

$$\dot{\Psi}_4 = -\Psi_2, \tag{3.41}$$

$$\dot{\Psi}_5 = \frac{c\beta}{m^2} k\omega.$$

for all three cases.

For determination of the optimal controls in all three cases (a), (b), (c), we will have the same conditions:

$$\beta = \bar{\beta} \text{ for } k_\beta > 0,$$

$$\beta = 0 \text{ for } k_\beta < 0, \tag{3.42}$$

$$\text{tg } \omega = \frac{\Psi_4}{\Psi_3} \text{ for } k_\omega = +\sqrt{\Psi_3^2 + \Psi_4^2}\,.$$

We write the transversality conditions as

(a):

$$\left[-\delta m - H\delta t + \Psi_1\delta x + \Psi_2\delta y + \Psi_3\delta u + \Psi_4\delta v + \Psi_5\delta m\right]_0^T = 0, \tag{3.43}$$

(b):

$$\left[\delta t - H\delta t + \Psi_1\delta x + \Psi_2\delta y + \Psi_3\delta u + \Psi_4\delta v + \Psi_5\delta m\right]_0^T = 0, \tag{3.44}$$

(c):

$$\left[\frac{\partial R}{\partial m}\delta m + \frac{\partial R}{\partial t}\delta t - H\delta t + \Psi_1\delta x + \Psi_2\delta y + \Psi_3\delta u + \Psi_4\delta v + \Psi_5\delta m\right]_0^T = 0. \tag{3.45}$$

According to the boundary conditions (3.33), (3.34), the expressions (3.43) - (3.45) are reduced, respectively, to (a):

$$\Psi_3(T) = 0,$$

$$\Psi_4(t) = 0,$$

$$\Psi_5(T) = 1, \tag{3.46}$$

$$H(T) = \Psi_1(T)u_T + \Psi_2(T)v_T - \beta(T) = 0;$$

(b):

$$\Psi_3(T) = 0,$$

$$\Psi_4(T) = 0,$$

$$\Psi_5(T) = 0, \tag{3.47}$$

$$H(T) = \Psi_1(T)u_T + \Psi_2(T)v_T = 1;$$

(c):

$$\Psi_3(T) = 0,$$

$$\Psi_4(t) = 0,$$

$$\Psi_5(T) = -\left(\frac{\partial R}{\partial m}\right)_T, \tag{3.48}$$

$$H(T) = \Psi_1(T)u_T + \Psi_2(T)v_T + \left(\frac{\partial R}{\partial m}\right)_T \beta(T) = -\left(\frac{\partial R}{\partial t}\right)_T .$$

The first four equations of (3.41) give

$$\Psi_1 = c_1, \qquad \Psi_2 = c_2,$$
$$\Psi_3 = c_3 - c_1 t, \quad \Psi_4 = c_4 - c_2 t, \tag{3.49}$$

and, since $\Psi_3(T) = \Psi_4(T) = 0$ in all cases, it is possible to write

$$\Psi_3 = c_1(T-t), \qquad \Psi_4 = (c_2(T-1) \tag{3.50}$$

By virtue of (3.50), it is possible to conclude that in all three cases the aircraft proceeds so that the thrust vector is at a constant angle with the horizontal, i.e.

$$\mathrm{tg}\ \omega = \frac{c_2}{c_1} = \mathrm{const.} \tag{3.51}$$

With the help of (3.39), (3.41), it is possible to write the equation

$$\dot{k}_\beta = \frac{c}{m}\dot{k}_\omega. \tag{3.52}$$

By (3.50), we have

$$k_\omega = +\sqrt{\Psi_3^2 + \Psi_4^2} = (T-t)\sqrt{c_1^2 + c_2^2}\ , \tag{3.53}$$

Eq. (3.52) is re-written in the form

$$\dot{k}_\beta = \frac{c}{m}\sqrt{c_1^2 + c_2^2}\ . \tag{3.54}$$

In view of the fact that the function m(t) is always positive, it follows that k_β is a monotonically decreasing function for any $0 \leq \beta \leq \bar{\beta}$ and the law (3.42) may have only one switching point.

The following fuel expenditure program is obvious:

(a):

$$\beta = \bar{\beta}, \quad 0 \le t \le t^*_{(\alpha)},$$
$$\beta = 0, \quad t^*_{(\alpha)} \le t \le T_{(\alpha)}. \tag{3.55}$$

(b):

$$\beta = \bar{\beta}, \quad 0 \le t \le T_{(b)} = t^*_{(b)}. \tag{3.56}$$

Thus, the function $k_\beta(T) = 0$ by virtue of (3.47) and it follows that it does not change sign in $0 \le t \le T$.

(c):

$$\beta = \bar{\beta}, \quad 0 \le t \le t^*_{(c)},$$
$$\beta = 0, \quad t^*_{(c)} \le t \le T_{(c)}. \tag{3.57}$$

We integrate Eq. (3.30) under the initial conditions (3.33) and $\beta = \bar{\beta}$, i.e. during the time $0 \le t \le t^*$

$$m = m_0 - \bar{\beta}T,$$

$$u = c \cdot \cos\omega \quad \ln \frac{m_0}{m_0 - \bar{\beta}t},$$

$$v = -gt + c \cdot \sin\omega \cdot \ln \frac{m_0}{m_0 - \bar{\beta}t}, \tag{3.58}$$

$$x = \frac{c \cdot \cos\omega}{\bar{\beta}} \left[m_0 + (m_0 - \bar{\beta}t)\left(\ln \frac{m_0 - \bar{\beta}t}{m_0} - 1 \right) \right],$$

$$y = \frac{c \cdot \sin\omega}{\bar{\beta}} \left[m_0 + (m_0 - \bar{\beta}t)\left(\ln \frac{m_0 - \bar{\beta}t}{m_0} - 1 \right) - \frac{gt^2}{2} \right].$$

At t^*, the values of the variable will then be used as the initial values for continuing the integration for $t \in [t^*, T]$. The latter imposes a continuity condition upon the trajectories of x, y, u, v, m. Thus, we have the following integrals:

$$
\begin{aligned}
m &= m_T, \\
u &= u(t^*), \\
v &= v(t^*) - gt, \\
x &= x(t^*) + u(t^*)t, \\
y &= y(t^*) + v(t^*)t - \frac{gt^2}{2}, \qquad t \in \left[t^*, T\right].
\end{aligned}
\tag{3.59}
$$

For obtaining the final solution, it remains to determine the constants c_1, c_2 and the unknown t^*, T.

For this we have

(a):

$$
\begin{aligned}
&k_\beta(t^*_{(\alpha)}) = 0 \\
&H(T_{(\alpha)}) = c_1^{(\alpha)} u(T_{(\alpha)}) + c_2^{(\alpha)} v(T_{(\alpha)}) = 0, \\
&x(T_{(\alpha)}) = x(t^*_{(\alpha)}) + u(t^*_{(\alpha)})(T_{(\alpha)} - t^*_{(\alpha)}) = x_T, \\
&y(T_{(\alpha)}) = y(t^*_{(\alpha)}) + v(t^*_{(\alpha)})(T_{(\alpha)} - t^*_{(\alpha)}) - \frac{g}{2}(T_{(\alpha)} - t^*_{(\alpha)})^2 = 0.
\end{aligned}
\tag{3.60}
$$

(b):

$$T_{(b)} = t^*_{(b)},$$

$$H(T_{(b)}) = c_1^{(b)} u(T_{(b)}) + c_2^{(b)} v(T_{(b)}) = 1,$$

$$x(T_{(b)}) = x(t^*_{(b)}) + u(t^*_{(b)})(T_{(b)}-t^*_{(b)}) = x_T, \tag{3.61}$$

$$y(T_{(b)}) = y(t^*_{(b)}) + v(t^*_{(b)})(T_{(b)}-t^*_{(b)}) - \frac{g}{2}(T_{(b)}-t^*_{(b)})^2 = 0.$$

It is assumed that the solution of the equations (3.60), (3.61) exists. It is not difficult to prove that this is indeed the case.

Now we consider the analogous equations in problem (c). We have

$$\begin{aligned}
&(c):\\
&k_\beta(t^*_{(c)}) = 0,\\
&c_1^{(c)}u(T_{(c)}) + c_2^{(c)}v(T_{(c)}) = -\left(\frac{\partial R}{\partial t}\right)_{T_{(c)}},\\
&x(t^*_{(c)}) + u(t^*_{(c)})(T_{(c)}-t^*_{(c)}) = x_T,\\
&y(t^*_{(c)}) + v(t^*_{(c)})(T_{(c)}-t^*_{(c)}) - \frac{g}{2}(T_{(c)}-t^*_{(c)})^2 = 0.
\end{aligned} \tag{3.62}$$

In the system of equations (3.60), (3.61), (3.62), the last two equations coincide. Their solutions must satisfy the first two equations. This may always be arranged for any continuous, positive-definite function R.

5. On Sufficient Conditions for Problems of Scalar Optimization

As is known (150), sufficient conditions for optimality in the problem of Mayer for fixed α are the following:

1. The function

$$\tilde{H}\left(x,u,\frac{\partial\Phi_\alpha}{\partial x}\right) = \sum_{i=1}^{n} f_i(x,u)\frac{\partial\Phi_\alpha}{\partial x_i} \tag{3.63}$$

must have an absolute minimum in the region $N(x,u) \geq 0$ with respect to u, which determines

$$u^{(\alpha)} = u^{(\alpha)}\left(x, \frac{\partial \Phi_\alpha}{\partial x}\right). \tag{3.64}$$

2. Under condition 1, the equation

$$\sum_{i=1}^{n} \frac{\partial \Phi_\alpha}{\partial x_i} f_i\left(x, u^{(\alpha)}\left(x, \frac{\partial \Phi_\alpha}{\partial x}\right)\right) = 0 \tag{3.65}$$

must have a smooth solution satisfying the boundary condition

$$\Phi_\alpha(x(t_\alpha)) = G_\alpha(x(T_\alpha)). \tag{3.66}$$

It is assumed that the solutions exist for all α.

We note that sufficient conditions for optimality in the problem of Mayer may be formulated in an analogous form following the works (151-153).

6. On Sufficient Conditions for Problems of Vector Optimization

For the vector optimization problems, sufficient conditions reduce to the following:

1. There must exist an absolute minimum for the function

$$\tilde{H}\left(x, u, \frac{\partial \Phi}{\partial x}\right) = \sum_{i=1}^{n} f_i(x,u) \frac{\partial \Phi}{\partial x_i} \tag{3.67}$$

with respect to u in the region $N(x,u) \geq 0$, which determines

$$u^0 = u^0\left(x, \frac{\partial \Phi}{\partial x}\right). \tag{3.68}$$

2. Under condition 1, there must exist a smooth solution of the equation

$$\sum_{i=1}^{n} \frac{\partial \Phi}{\partial x_i} f_i\left(x, u^0\left(x, \frac{\partial \Phi}{\partial x}\right)\right) = 0, \tag{3.69}$$

which must satisfy the boundary conditions

$$\Phi(x(T)) = R(y(T)). \tag{3.70}$$

We compare formulas (3.63) - (3.66) with the formulas (3.67) - (3.70). Whenever the functions (3.63) and (3.67) are identical in form, then the structures of $u^{(\alpha)}$ and $u^{(0)}$ will also be the same. This also follows from the Pontryagin maximum principle (6). This means that the control u providing the minimum of the functions $\tilde{H}$, H_α is invariant relative to the form of the optimization functional, and the functions $\tilde{H}$, $\tilde{H}_\alpha$ take on their minimum at the same point of the space, $u(x, \frac{\partial \Phi}{\partial x})$.

Now the following questions arise: does there exist a smooth function $\Phi(x_1, x_2, \ldots, x_n)$ which simultaneously satisfies (3.69) and (3.70), if it is known that there exist smooth functions $\Phi_1(x_1, \ldots, x_n), \Phi_2(x_1, \ldots, x_n), \ldots, \Phi_k(x_1, \ldots, x_n)$ satisfying (3.65) and (3.66)? If yes, how should we construct the function $\Phi(x_1, \ldots, x_n)$ from the functions $\Phi_1, \Phi_2, \ldots, \Phi_k$?

Several cases of construction of the function $\Phi(x_1, x_2, \ldots, x_n)$ will be considered in Chapter V in connection with the problem of analytic construction of regulators by dynamic programming.

CHAPTER IV
PROGRAMMING OPTIMAL TRAJECTORIES FOR PROBLEMS WITH VECTOR-VALUED CRITERIA

1. Introduction

As is well known, the solution of the problem of programming optimal trajectories is based on the mathematical apparatus of the calculus of variations--the Pontryagin maximum principle (6). This method is a general method for the solution of variational problems, defining ordinary differential equations for piecewise-continuous admissible controls. By obtaining the solutions of particular variational problems arising in modern technology, the maximum principle has become an effective tool in the hands of engineers for design and computation of modern automatic control systems.

Following (18), we select a statement of the variational problem in the Mayer form. The reason for this, as noted in (18), is the uniform structure of the Mayer problem and the possibility of standardization of the mathematical constructions, which is very important for extensions.

On the basis of the maximum principle formulated for the problem of Mayer, we give a solution of the problem of programming optimal trajectories for vector-valued performance indices.

The current questions were first considered in the works (142, 144,146,147).

2. A General Approach to the Solution of Programming Problems

Let the motion of the controlled system be described by the vector differential equation (2.2).

We consider the problem of optimizing the vector functional (2.3) with components

$$
\begin{aligned}
I_1(u) &= \int_{t_0}^{T} \omega_1(x(t),u(t))dt,\\
&\cdots\cdots\cdots\cdots\cdots\\
I_r(u) &= \int_{t_0}^{T} \omega_r(x(t),u(t))dt, \qquad (4.1)\\
I_{r+1}(u) &= \omega_{r+1}(x(T)),\\
&\cdots\cdots\cdots\cdots\cdots\\
I_k(u) &= \omega_k(x(T)).
\end{aligned}
$$

The integral of the first r functionals is positive definite and continuously differentiable in its arguments. The initial time t_0 is given, while the final time T in each subproblem is free since the different functionals take on their optima for different T.

We consider optimization of the system described by Eq, (2.2) under the boundary conditions

$$
\begin{aligned}
x_i(t_0) &= x_{i0} \quad (i=1,\ldots,n),\\
x_i(T) &= x_{iT} \quad (i=1,\ldots,\beta<n),
\end{aligned} \qquad (4.2)
$$

i.e. the values $x_{\beta+1}(T),\ldots,x_n(T)$ are free and we have the constraint

$$|u_i| \leq \bar{u}_i, \quad (i=1,\ldots,m). \tag{4.3}$$

The conditions imposed on (2.2) (cf. section 2, Chapter II) guarantee the existence of optimal controls $u^{(1)}, u^{(2)}, \ldots, u^{(k)}$ for the individual functionals (4.1) and the optimal trajectories corresponding to the controls (under independent optimization with respect to each individual functional). They, in turn, determine the optimal values of the corresponding functionals which we denote by

$$I_\alpha(u^{(\alpha)}) = I_{\alpha 0} \quad (\alpha = 1,\ldots,k). \tag{4.4}$$

The optimization of the separate functionals (4.1) will, generally speaking, be carried out over different time intervals $T_1, T_2, \ldots, T_k$. The quantities $T_{\alpha 0}$ $(\alpha=1,\ldots,k)$ are quite definite numbers and in the sequel we may assume that they are given. Now, according to section 2, Chapter II, optimization of the vector functional (2.3) with components (4.1) is reduced to minimization of the sum

$$R(u) = \sum_{\alpha=1}^{r} \left[\int_{t_0}^{T} \omega_\alpha(x(t),u(t))dt - I_{\alpha 0}\right]^2 + \sum_{\alpha=r+1}^{k} \left[\omega_\alpha(x(T) - I_{\alpha 0})\right]^2. \tag{4.5}$$

The problem of minimizing the functional (4.5) under the dynamics (2.2) is an ordinary variational problem which may be solved relatively simply if we represent it in the equivalent Mayer form. With this goal in mind, we introduce the new variables

$$y_\alpha(t) = \int_{t_0}^{t} \omega_\alpha(x(t),u(t))dt \quad (\alpha=1,\ldots,r), \tag{4.6}$$

which are the solutions of the differential equations

$$\dot{y}_\alpha = \omega_\alpha(x(t),u(t)) \quad (\alpha=1,\ldots,r) \tag{4.7}$$

under the initial conditions

$$y_\alpha(t_0) = 0 \quad (\alpha=1,\ldots,r. \tag{4.8}$$

These equations are adjoined to Eqs. (2.2), along with adding condition (4.8) to condition (4.2). Using the notation (4.7), the functional (4.5) is written as

$$R(u) = \sum_{\alpha=1}^{r} \left[y_\alpha(T)-I_{\alpha 0}\right]^2 + \sum_{\alpha=r+1}^{k} \left[\omega_\alpha(x(T))-I_{\alpha 0}\right]^2. \tag{4.9}$$

As noted already, the value of T is not fixed. For convenience, we will minimize one-half the norm of (4.9), which does not change the problem in any way. We denote the time-varying expression (4.9) as

$$2G(x(t),y(t)) = \sum_{\alpha=1}^{r} \left[y_\alpha(t)-I_{\alpha 0}\right]^2 + \sum_{\alpha=r+1}^{k} \left[\omega_\alpha(x(t))-I_{\alpha 0}\right]^2. \tag{4.10}$$

Now it is possible to formulate an equivalent problem of Mayer: among the admissible curves x(t), y(t), u(t), satisfying the conditions (2.2), (4.2), (4.3), (4.7) and (4.8), find those which minimize the functional

$$\Delta G = G\Big|_{t_0}^{T} = G(T) - G(t_0) = \tfrac{1}{2}R(u) - \tfrac{1}{2}\sum_{\alpha=1}^{r} I_{\alpha 0}^2 - \tfrac{1}{2}\sum_{\alpha=r+1}^{k} \left[\omega_\alpha(x(t_0))-I_{\alpha 0}\right]^2. \tag{4.11}$$

It is clear that the control u^0 minimizing ΔG also minimizes R(u).

The solution of the Mayer problem may be carried out using the maximum principle in a form tailored for the problem of Mayer (18): in order that the control u(t) provide a strong minimum for the functional (4.11) under the relations (2.2), (4.7) and the boundary conditions (4.2), (4.8), it is necessary that there exists a non-zero vector $\Psi(\Psi_1,\ldots,\Psi_{n+r})$, defined by the equations

$$\dot{\Psi}_i = -\frac{\partial H}{\partial x_i} \quad (i=1,\ldots,n),$$
$$\dot{\Psi}_{n+\alpha} = -\frac{\partial H}{\partial y_\alpha} \quad (\alpha=1,\ldots,r), \tag{4.12}$$

under which the function

$$H = \sum_{i=1}^{n} \Psi_i f_i(x,u,t) + \sum_{\alpha=1}^{r} \Psi_{n+\alpha}\omega_\alpha(x,u) \tag{4.13}$$

assumes its maximum at u and the transversality condition

$$\left| \delta G - H\delta t + \sum_{i=1}^{n} \Psi_i \delta x_i + \sum_{\alpha=1}^{r} \Psi_{n+\alpha}\delta y_\alpha \right|_{t_0}^{T} = 0. \tag{4.14}$$

is satisfied. The condition that (4.13) be maximized defines u^0 as a function

$$u^0 = u^0(x,\Psi,t). \tag{4.15}$$

Equation (4.12) and the expression (4.13) determine the system of equations for the vector Ψ:

$$\dot{\Psi}_j = -\sum_{i=i}^{n} \Psi_i \frac{\partial f_i}{\partial x_j} - \sum_{\alpha=1}^{r} \Psi_{n+\alpha} \frac{\partial \omega_\alpha}{\partial x_j} \quad (j=1,\ldots,n),$$
$$\dot{\Psi}_{n+\alpha} = 0 \quad (\alpha=1,\ldots,r), \tag{4.16}$$

the right part of which will contain only the variables x, Ψ, t.

Equations (2.2), (4.7) and (4.16) define a system of 2(n+r) differential equations for the variables

$$x_1,\ldots,x_n, \quad y_1,\ldots,y_r, \quad \Psi_1,\ldots,\Psi_{n+r}.$$

For the solution of the given system, we have the n + r initial conditions

$$x_i(t_0) = x_{i0} \quad (i=1,\ldots,n),$$
$$y_\alpha(t_0) = 0 \quad (\alpha=1,\ldots,r) \tag{4.17}$$

and the β conditions at the right end of the trajectory

$$x_i(T) = x_{iT} \quad (i=1,\ldots,\beta<n). \tag{4.18}$$

For solution we require (2n+2r+1) conditions, since we have a definite value of T; (n-β+r+1) conditions will not suffice.

Since the initial points of the trajectories of x(t), y(t) and the final point of the trajectory x(t) are fixed, the transversality condition (4.14) splits into (n-β+r+1) equations

$$H(T) = 0,$$
$$y_\alpha(T) - I_{\alpha 0} + \Psi_{n+\alpha}(T) = 0 \quad (\alpha=1,\ldots,r), \tag{4.19}$$
$$\sum_{\alpha=r+1}^{k}\left[\omega_\alpha(x(T))-I_{\alpha 0}\right]\left(\frac{\partial\omega_\alpha}{\partial x_i}\right)_{t=T} + \Psi_i(T) = 0 \quad (i=\beta+1,\ldots,n),$$

connecting the values x(T), y(T), Ψ(T), T.

Thus, for determining the control (4.15) which is optimal with respect to the vector functional with components (4.1), it remains to integrate the system (2.2), (4.7), (4.16) under the boundary conditions (4.17) - (4.19) and to substitute the result into the expression (4.15).

Consequently, the formulated problem on optimization of the vector functional (2.3) with components (4.1) is reduced to the solution of a two-point boundary value problem for Eqs. (2.2), (4.7) and (4.16) under the conditions (4.17) - (4.19). We illustrate the solution of such a two-point boundary value problem on a concrete example.

3. A Simple Second-Order Example

We consider the control object which, in dimensionless time, is described by the system of equations

$$\dot{x}_1 = x_2, \quad \dot{x}_2 = u, \tag{4.20}$$

defined in the closed region

$$\chi(x) = 3 - |x_1| \geq 0. \tag{4.21}$$

Here x_1, x_2 are state variables and u is the control variable.

From the class of admissible controls, defined by the constraint,

$$|u| \leq 1, \tag{4.22}$$

we look for a control $u^0(t)$ which transfers the system from the initial state

$$x_1(0) = 1, \quad x_2(0) = 0 \tag{4.23}$$

to the state

$$x_1(T) = 0 \tag{4.24}$$

while simultaneously minimizing the functional

$$I_1(u) = \int_0^T dt = T \tag{4.25}$$

and maximizing the value $x_2(T)$, i.e.

$$I_2(u) = x_2(T). \tag{4.26}$$

The problem of minimization of the functional (4.25) for such a system is considered in (6) and the control is defined as

$$u^{(1)} = \begin{cases} +1 \text{ for } x_1 < 0, \\ -1 \text{ for } x_1 > 0. \end{cases} \tag{4.27}$$

For the concrete initial condition (4.23), we will have

$$u^{(1)} = -1, \tag{4.28}$$

which, along the trajectory (Fig. 7, segment ABCD)

$$x_1 = -\frac{t^2}{2} + 1, \qquad x_2 = -t \tag{4.29}$$

transfers the system from the point (4.23) to the x_2-axis in the minimal time

$$I_{10} = T^{(1)} = \sqrt{2} \approx 1.414. \tag{4.30}$$

In this case, the quantity $x_2(T)$ takes on the value

$$I_2 = x_2^{(1)}(T) = -\sqrt{2} \approx -1.414. \tag{4.31}$$

Now we solve the problem of maximizing the functional (4.26) for a free time T. Without the constraint (4.21), such a problem makes no sense since, in the opposite case, it is possible to make the functional (4.26) infinite on an infinite-time interval, as there is no constraint on the control resources, which is usually contained in the equations (4.20).

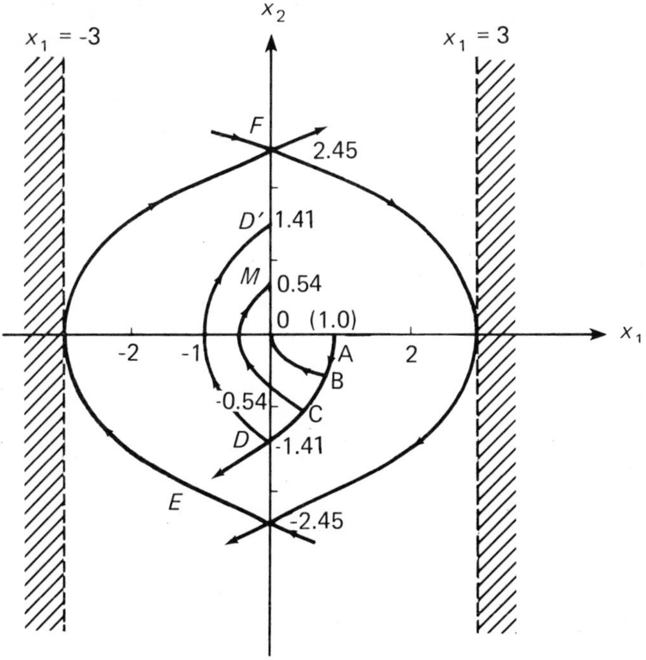

Fig. 7

This problem is formulated as a problem of Mayer in the following way: under the constraints (4.21), (4.22), we seek a control $u^{(2)}$, which transfers the system (4.20) from the position (4.23) to the position (4.24) and for which the functional

$$\Delta G = G\Big|_0^T = -x_2(T) \tag{4.32}$$

is minimized.

According to the maximum principle, the function $\tilde{H}$ will have the form ((10), Ch. IV)

$$\tilde{H} = \Psi_1 x_2 + \Psi_2 u + \mu\left(\frac{\partial\chi}{\partial x_1}\dot{x}_1 - \frac{\partial\chi}{\partial x_2}\dot{x}_2\right), \tag{4.33}$$

from which we have the control $u^{(2)}$ as

$$u^{(2)} = \text{sign } \Psi_2(t). \tag{4.34}$$

For determination of the function $\Psi_2(t)$ we have the equation

$$\begin{aligned} \dot{\Psi}_1 &= 0, \\ \dot{\Psi}_2 &= -\Psi_1 + \mu x_2 \cdot \text{sign } x_1, \end{aligned} \tag{4.35}$$

which we solve together with Eq. (4.20) under the boundary condition

$$\begin{aligned} &\chi(x(\tau)) = 3 - |x_1(\tau)| = 0, \\ &\left(\frac{\partial\chi}{\partial x_1}\dot{x}_1 + \frac{\partial\chi}{\partial x_2}\dot{x}_2\right)_{t=\tau} = -\frac{\partial|x_1|}{\partial x_1}x_2(\tau) = 0 \end{aligned} \tag{4.36}$$

and the transversality condition

$$|\delta G - \tilde{H}\delta t + \Psi_1\delta x_1 + \Psi_2\delta x_2|_0^T = 0. \tag{4.37}$$

The multiplier μ differs from zero at $t = \tau$, where τ is the moment at which the trajectory $x(t)$ reaches the boundary (4.21). With regard to conditions (4.23), (4.24), the last equation splits into the relations

$$\begin{aligned} &\Psi_2(T) = 1, \\ &\tilde{H}(T) = \Psi_1(T)x_2(T) + u(T) + \mu\left(\frac{\partial\chi}{\partial x_1}\dot{x}_1 + \frac{\partial\chi}{\partial x_2}\dot{x}_2\right)_{t=T} = 0. \end{aligned} \tag{4.38}$$

Thus, for solving the system (4.20), (4.35) and determination of T, μ, τ, we have the three boundary conditions (4.23), (4.24), the relations (4.38) and the condition (4.36).

Eqs. (4.35) define $\Psi_2(t)$ as a linear function which may change sign only once. The first of the relations (4.38) shows that if $\Psi_2(t)$ changes sign in the interval $0 \le t \le T$, then it must be an increasing function. Therefore, the optimizing control must have the form

$$u^{(2)} = \begin{cases} -1 \text{ for } 0 \le t \le t_{sw} \\ +1 \text{ for } t_{sw} \le t \le T. \end{cases} \tag{4.39}$$

Here t_{sw} denotes the switching time.

The solution of the system (4.20), (4.35) under the conditions (4.23), (4.36), (4.38), taking account of (4.39), shows that the trajectory of the system will coincide with the trajectory of (4.29) for $0 \le t \le t_{sw}$, while for $t_{sw} \le t \le T$ it will have the form

$$\begin{aligned} x_1 &= \frac{t^2}{2} - 4t + 5, \\ x_2 &= t - 4, \end{aligned} \tag{4.40}$$

which represents a parabolic equation passing through the point $x_1 = -3$, $x_2 = 0$ (Fig. 7, segment ABCDEF).

Thus, we determine the values

$$t_{sw} = 2, \tag{4.41}$$

$$I_1 = T^{(2)} = 4 + \sqrt{6} \simeq 6.449, \tag{4.42}$$

$$I_{20} = x_2^{(2)}(T) = \sqrt{6} \simeq 2.449. \tag{4.43}$$

Now, having the values I_{10}, I_{20} as in (4.30) and (4.43), we solve the problem of optimizing the functionals (4.25) and (4.26) simultaneously.

According to section 2, the minimization is for the sum

$$R(u) = \left[\int_0^T dt - 1.414\right]^2 + \left[2.449 - x_2(T)\right]^2 \tag{4.44}$$

for a free termination time T.

We introduce the notation

$$x_3(t) = \int_0^t dt = t \tag{4.45}$$

and write the differential equation

$$\dot{x}_3 = 1 \tag{4.46}$$

with the initial condition

$$x_3(0) = 0. \tag{4.47}$$

We denote the function (4.44) as a time-varying quantity as

$$2G(x_2(t), x_3(t)) = \left[x_3(t) - 1.414\right]^2 + \left[2.449 - x_2(t)\right]^2. \tag{4.48}$$

We formulate the given variational problem as a problem of Mayer: among the admissible curves $x_1(t)$, $x_2(t)$, $x_3(t)$, $u(t)$ satisfying the conditions (4.20)-(4.24), (4.46) and (4.47), find those which minimize the functional

$$\Delta G = G\Big|_0^T = G(T) - G(0) = \tfrac{1}{2}\left[x_3(T) - 1.414\right]^2 + \tfrac{1}{2}\left[2.449 - x_2(T)\right]^2 - \tfrac{1}{2}(1.414^2 + 2.449^2). \tag{4.49}$$

For the system of equations (4.20), (4.46), the function H is written in the form

$$H = \Psi_1 x_2 + \Psi_2 u + \Psi_3, \tag{4.50}$$

which, taken account of (4.42), determines the optimal control as

$$u^0 = \operatorname{sign} \Psi_2(t). \tag{4.51}$$

For determining the non-zero function $\Psi_2(t)$, we will have the system of differential equations

$$\dot{\Psi}_1 = 0, \quad \dot{\Psi}_2 = -\Psi_1, \quad \dot{\Psi}_3 = 0, \tag{4.52}$$

which must be solved, together with the system

$$\dot{x}_1 = x_2, \quad \dot{x}_2 = u, \quad \dot{x}_3 = 1 \tag{4.53}$$

under the boundary conditions

$$\begin{aligned} &x_1(0) = 1, \quad x_1(T^0) = 0 \\ &x_2(0) = 0, \\ &x_3(0) = 0, \end{aligned} \tag{4.54}$$

using (4.51) and the transversality condition

$$\left| \delta G - H\delta t + \Psi_1 \delta x_1 + \Psi_2 \delta x_2 + \Psi_3 \delta x_3 \right|_0^{T_0} = 0. \tag{4.55}$$

The transversality condition (4.55) gives the following identities:

$$H(T^0) = \Psi_1(T^0) x_2(T^0) + \Psi_2(T^0) u(T^0) + \Psi_3(T^0) = 0, \tag{4.56}$$

$$x_3(T^0) - 1.414 + \Psi_3(T^0) = 0, \tag{4.57}$$

$$x_2(T^0) - 2.449 + \Psi_2(T^0) = 0, \tag{4.58}$$

which, together with the boundary conditions (4.54), guarantee the solvability of the system (4.52), (4.53) and determination of a finite termination time T^0.

Just as in the case of optimization of the single functional I_2, the optimizing control is completely determined by the function $\Psi_2(t)$, which is a linear function of t (this comes from the system (4.52)). If this function has a zero in the interval $0 \leq t \leq T^0$, then it will be increasing. The latter assertion comes out of the condition (4.48) which is superimposed upon the system (4.52), (4.53) and from which we see that

$$\Psi_2(T^0) = 2.449 - x_2(T^0) \geq 0. \tag{4.59}$$

The inequality (4.59) occurs unconditionally since the possible maximal values of $x_2(T)$ were determined according to (4.43).

Thus, the control $u^0(t)$ optimizing the vector functional with components (4.25), (4.26) will have the form:

$$u^0(t) = \begin{cases} -1 \text{ for } 0 \leq t \leq t^0_{sw} \\ +1 \text{ for } t^0_{sw} \leq t \leq T^0, \end{cases} \tag{4.60}$$

where t_{sw} is the switching time and is subject to the condition

$$t^0_{sw} \leq t_{sw}. \tag{4.61}$$

The value of t_{sw} is determined by (4.41).

It remains to determine the values of t^0_{sw}, T^0 after which the stated vector optimization problem can be assumed to be solved.

The case of equality $t^0_{sw} = T^0$ corresponds to the minimal time system when the value of the functional (4.26) is free. As was noted above, in such a case the system's optimal regime has no switching.

The equality $t^0_{sw} = t_{sw}$ corresponds to the case of maximizing the functional I_2 for a free value of T.

Now we determine the values t^0_{sw}, T^0.

The transient process of the system under the optimal control (4.60) will have the form (Fig. 7, trajectory ABCM):

$$\left.\begin{aligned} x_1 &= -\frac{t^2}{2} + 1 \\ x_2 &= -t \end{aligned}\right\} \text{for } 0 \leq t \leq t^0_{sw}, \tag{4.62}$$

$$\left.\begin{aligned} x_1 &= \frac{t^2}{2} + At + B \\ x_2 &= t + A \end{aligned}\right\} \text{for } t^0_{sw} \leq t \leq T^0, \tag{4.63}$$

$$x_3 = t \text{ for } 0 \leq t \leq T^0. \tag{4.64}$$

According to the boundary conditions (4.54), we will have

$$x_1(T^0) = \frac{(T^0)^2}{2} + AT^0 + B = 0, \tag{4.65}$$

$$x_2(T^0) = T^0 + A, \tag{4.66}$$

$$x_3(T^0) = T^0. \tag{4.67}$$

From the continuity conditions on the functions $x_1(t)$ and $x_2(t)$, we determine

$$A = -2\, t^0_{sw}, \tag{4.68}$$

$$B = t^{0^2}_{sw} + 1. \tag{4.69}$$

The solution of the system (4.52) has the form

$$\Psi_1(t) = \Psi_1(T^0) = \text{const}, \tag{4.70}$$

$$\Psi_2(t) = -\Psi_1(T^0)t + K, \tag{4.71}$$

$$\Psi_3(t) = \Psi_3(T^0) = \text{const}. \tag{4.72}$$

The fact that $\Psi_2(t)$ is zero at the switching time t_{sw} determines the constant of integration

$$K = \Psi_1(T^0)t^0_{sw}, \tag{4.73}$$

According to the expressions (4.57), (4.58), (4.66), (4.67), (4.71) and (4.73), we may write

$$\begin{aligned} \Psi_1(T^0) &= \frac{2.449 - T^0 - A}{t^0 - T^0}, \\ \Psi_2(T^0) &= 2.449 - T^0 - A, \\ \Psi_3(T^0) &= 1.414 - T^0. \end{aligned} \tag{4.74}$$

Using (4.68), (4.69), (4.74), the relations (4.56), (4.65) determine a system of second-order nonlinear algebraic equations for the unknowns t^0_{sw}, T^0 in the following form:

$$\begin{aligned} T^{0^2} - 1.414\,T^0 - 2t^{0^2}_{sw} - 1.035\,t^0_{sw} &= 0, \\ T^{0^2} - 4t^0_{sw}T^0 + 2t^{0^2}_{sw} + 2 &= 0. \end{aligned} \tag{4.75}$$

The equations obtained are subject to the three constraints

$$t^* \leq t_{sw}^0 \leq t_{sw} = 2,$$

$$1,414 = T^{(1)} \leq T^0 \leq T^{(2)} = 6.449,$$

where t^* is determined from the relations

$$x_1 = \frac{t^2}{2} - 2t^*t + t^{*2} + 1 = 0,$$
$$x_2 = t - 2t^* = 0 \tag{4.76}$$

which gives $t^* = 1$.

Conditions (4.76) determine the limiting minimal value of the switching time. For $t_{sw}^0 < t^*$, the boundary conditions (4.54) will not be valid, since the parabola (4.63), (4.68), (4.69) does not intersect the axis $x_1 = 0$.

As a result of the solution of the system of equations (4.75) in the region

$$1 \leq t_{sw}^0 \leq 2,$$

$$1.41 \leq T^0 \leq 6.44$$

we obtain the values

$$t_{sw}^0 = 1.0693,$$
$$T^0 = 2.6785. \tag{4.77}$$

Under the control (4.60), where t_{sw}^0 and T^0 are determined by (4.77), the value $x_2(T^0)$ is determined from (4.62) and (4.63) as

$$x_2(T^0) = T^0 - 2t_{sw}^0 = 0.5399.$$

TABLE 1.

Control Object and Boundary Conditions	Form of the Control	Functional to be Minimized
$\dot{x}_1 = x_2,\ \dot{x}_2 = u.$ $\lvert u\rvert \le 1,\ \lvert x_1\rvert \le 3.$ $x_1(0) = 1,\ x_1(T) = 0.$ $x_2(0) = 0.$	$u^{(1)} = -1$ $0 \le t \le \sqrt{2}$	$I_1 = \int_0^T dt$
$\dot{x}_1 = x_2,\ \dot{x}_2 = u.$ $\lvert u\rvert \le 1,\ \lvert x_1\rvert \le 3.$ $x_1(0) = 1,\ x_1(T) = 0.$ $x_2(0) = 0.$	$u^{(2)} = \begin{cases} -1, 0 \le t \le 2 \\ +1, 2 \le t \le 6.449 \end{cases}$	$-I_2 = -x_2(T)$
$\dot{x}_1 = x_2,\ \dot{x}_2 = u.$ $\lvert u\rvert \le 1,\ \lvert x_1\rvert \le 3.$ $x_1(0) = 1,\ x_1(T) = 0.$ $x_2(0) = 0.$	$\bar{u} = \begin{cases} -1, 0 \le t \le 1 \\ +1, 1 \le t \le 2 \end{cases}$	$I = I_1 - I_2 = \int_0^T dt - x_2(T)$
$\dot{x}_1 = x_2,\ \dot{x}_2 = u.$ $\lvert u\rvert \le 1,\ \lvert x_1\rvert \le 3.$ $x_1(0) = 1,\ x_1(T) = 0.$ $x_2(0) = 0.$	$u^0 = \begin{cases} -1, 0 \le t \le 1.0693 \\ +1, 1.0693 \le t \le 2.6785 \end{cases}$	$I = \left[\int_0^T dt - 1.414\right]^2 + \left[2.449 - x_2(T)\right]^2$

Form of the Trajectory	Functional Value $I_1=\int_0^T dt$	Functional Value $I_2=x_2(T)$
$x_1 = \frac{t^2}{2} + 1$ $x_2 = -t$ $0 \leq t \leq \sqrt{2}$	$I_{10}=\sqrt{2}\approx 1.414$	$I_2^{(1)} = -\sqrt{2} \approx$ ≈ -1.414
$x_1 = -\frac{t^2}{2} + 1$ $x_2 = -t$ $0 \leq t \leq 2;$ $x_1 = \frac{t^2}{2} - 4t + 5$ $x_2 = t - 4$ $2 \leq t \leq 6.449.$	$I_1^{(2)} = 4 + \sqrt{6}$ ≈ 6.449	$I_{20} = \sqrt{6} \approx 2.449$
$x_1 = -\frac{t^2}{2} + 1$ $x_2 = -t$ $0 \leq t \leq 1;$ $x_1 = \frac{t^2}{2} - 2t + 2$ $x_2 = t - 2$ $1 \leq t \leq 2$	2	0
$x_1 = -\frac{t^2}{2} + 1$ $x_2 = -t$ $0 \leq t \leq 1.0693;$ $x_1 = \frac{t^2}{2} - 2.1386t +$ $+ 2.1434$ $x_2 = t - 2.1386$ $1.0693 \leq t \leq 2.6785$	$T^0 = 2.6785$	$x_2^0(T) = 0.5399$

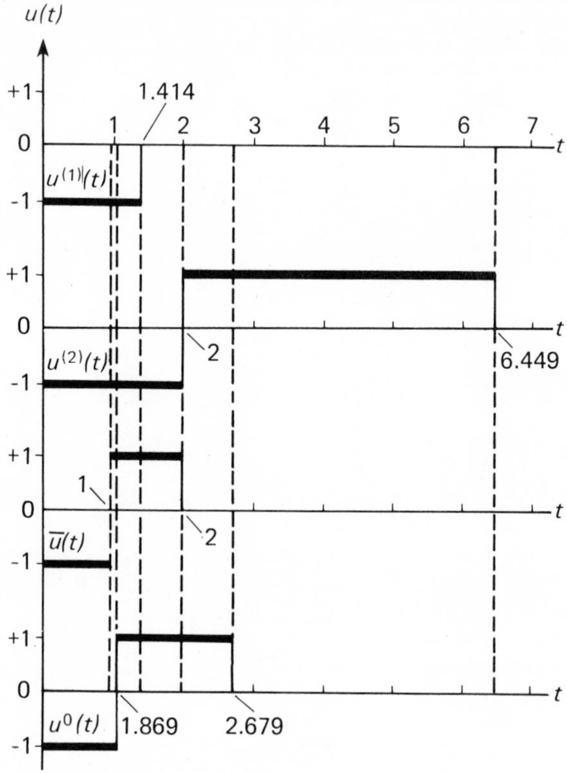

Fig. 8

Thus, using the rules (4.60), (4.77) for the control system (4.20), the components (4.25), (4.26) of the vector functional assume the values

$$I_1(u^0) = T^0 = 2.6785, \qquad I_2(u^0) = x_2(T^0) = 0.5399.$$

In Table 1 we present the results for comparing values of the functionals (4.25), (4.26) for the different control laws (4.28), (4.39), (4.60).

Here we have presented results for the solution of the stated problem for minimization of the algebraic sum of the functionals

$$I(u) = I_1 + I_2 = \int_0^T dt - x_2(T).$$

It is not difficult to calculate that in this case the control has the form

$$\bar{u} = \begin{cases} -1 \text{ for } 0 \leq t \leq 1 \\ +1 \text{ for } 1 \leq t \leq 2 \end{cases} \quad (4.78)$$

and the corresponding functionals (4.25), (4.26) assume the values

$$I_1 = T = 2, \quad I_2 = x_2(T) = 0.$$

As was shown, in the case given the preferred functional turned out to be the functional I_1. The corresponding extremal trajectory will be the segment ABO in Fig. 7.

A graphical representation of the control functions (4.28), (4.39), (4.60) and (4.78) is given in Fig. 8.

Under the control (4.60), which is optimal for the vector interior, the sum of the functionals $I_1 + (-I_2)$ has a small deviation

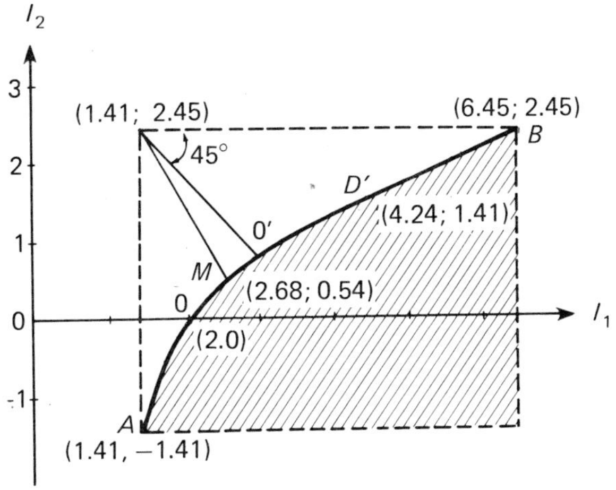

Fig. 9

from that value which is obtained under the control (4.78), but the individual values of the functionals I_1 and I_2 are found to be the best approximations to the values I_{10} and I_{20}.

The set of Pareto-optimal (non-improvable) solutions in this example is determined in the $\{I_1, I_2\}$ by the continuous, convex curve AB (Fig. 9). In order to convince ourselves of this, we write the equation (4.62) and (4.63) in the form

$$x_1(\tau) = 0.5\tau^2 - s\tau + 1 - 0.5s^2, \qquad x_2(\tau) = \tau - s,$$

in which $\tau = t - s$, where s denotes the moment when the control u(t) switches from -1 to +1. It is not difficult to see that attainment of the state $x_1(T) = 0$ is realized for values $s \in [1,2]$ and the number of switchings equals unity, with $x_2(T)$ assuming values from the interval $(-\sqrt{2}, \sqrt{6})$ (segment DF in Fig. 7). The value $x_2(T) = -\sqrt{2}$ is taken on without switching. With the help of the above-mentioned formulas for $x_1(\tau)$ and $x_2(\tau)$, we express the functionals I_1 and I_2 as functions of s.

We obtain

$$I_1(u) = T = 2s \pm \sqrt{2(s^2-1)},$$

$$I_2(u) = x_2(T) = \pm \sqrt{2(s^2-1)},$$

in which the sign "+" corresponds to the interval $s \in [1,2]$ for attaining non-negative values of $x_2(T)$, while the sign "-" corresponds to the interval $s \in (1, \sqrt{2})$ for attaining negative values of $x_2(T)$. (We note that in all the cases considered, the number of switchings is unity except when $x_2(T) = -\sqrt{2}$.) Now it is not difficult to see that for any $s \in [1,2]$, we have the inequalities

$$\frac{dI_2}{dI_1} = \frac{dI_2}{ds} \Big/ \frac{dI_1}{ds} > 0,$$

$$\frac{d^2I_2}{dI_1^2} = \frac{d}{ds}\left(\frac{dI_2}{dI_1}\right) \Big/ \frac{dI_1}{ds} < 0,$$

which is indicated on the continuous, convex curve AB in Fig. 9.

The current example was studied in (99) for cases when the functional approximations subject to minimization were expressed in the norms

$$R_L = \left[(T-\sqrt{2})^L + (\sqrt{6}-x_2(T))^L\right]^{1/L}, \quad L \in [1,\infty].$$

The Pareto-optimal points O, M, O' on the curve AB (Fig. 9) are determined by the values $L = 1,2,\infty$, respectively. The (utopian) point, having coordinates $(\sqrt{2},\sqrt{6})$ is joined by a line in the direction to the normal to the curve AB at the point M. The point O' is situated between the points M and D' (Fig. 9) where D' is a mirror image of the point D in the x-axis (Fig. 7). The line connecting the utopian point with the point O' forms an angle of 45° to the horizontal (Fig. 9). In its dependence on the parameter L, the Pareto-optimal solution changes so that the corresponding point moves continuously, and monotonically from the point O to the point O' (Figs. 7 and 9) under variation of L from 1 to ∞.

4. On the Problem of Optimal Flight to a Given Location

There may be different requirements for the launching of a rocket to a given location. For example, the process may be optimal if it is carried out in minimal time. Optimality may also be defined in terms of minimal fuel expenditure or the optimization requirement may involve the maximal velocity of the rocket (which determines its kinetic energy) at the required terminal point. We may enumerate similar requirements ad infinitum. They define the meaning of optimal flight control. It is natural that for optimization of the flight with respect to only one criterion other criterion are ignored and, in many cases, they are so bad that the optimal program of the rocket's flight may become unacceptable.

There arises the problem of simultaneous optimization of several aspects of the aircraft. Without fail, simultaneous optimization of several criteria will be accompanied by a degradation of each performance index taken separately, with respect to the value attained under optimization only with respect to each scalar criterion. Nevertheless, it turns out to be desirable as a program of optimal flight to choose a program which is based upon a compromise combination of values of all the required optimality criteria.

a. Minimal Fuel Expenditure Flight in a Vertical Plane. The motion of an aircraft in a vertical plane under specific conditions is described by a system of five differential equations (154)

$$
\begin{aligned}
\dot{x} &= v \cos \gamma, \\
\dot{h} &= v \sin \gamma, \\
\dot{v} &= -g \sin \gamma + \frac{v_E \beta}{m} \cos \varepsilon - \frac{D}{m}, \\
\dot{\gamma} &= -\frac{g}{v} \cos \gamma + \frac{v_E \beta}{m} \sin \varepsilon + \frac{L}{mv}, \\
\dot{m} &= -\beta.
\end{aligned}
\tag{4.79}
$$

Here x is the horizontal state, h is the altitude; v is the velocity; γ is the angle-of-attack of the trajectory with respect to the horizontal; m is the total mass; ε is the angle between the thrust and velocity vectors; β is the mass of fuel expended; v_E is the effective velocity of the outgoing combustion products; D(h,v) is the drag; L(h,v) is the lift and g is the gravitational constant (• denotes differentiation with respect to time).

The equations are defined in the region $N \geq 0$ consisting only of the constraint $0 \leq \beta \leq \bar{\beta}$.

The functions $\beta(t)$, $\varepsilon(t)$ are the controls determining the flight program of the aircraft described by Eqs. (4.79).

As the functional to be optimized, we take

$$\Delta G = G\Big|_0^T = -(m_T - m_0) = m_0 - m_T, \tag{4.80}$$

which represents the fuel expenditure in transferring the aircraft to the required location.

We assume that the admissible control $\beta \in \Omega$, which consists of the class of piecewise-continuous functions satisfying

$$0 \leq \beta \leq \bar{\beta}. \tag{4.81}$$

We are given the following fixed boundary conditions for the problem

$$x(0) = x_0, \quad h(0) = h_0, \quad v(0) = v_0, \quad \gamma(0) = \gamma_0, \quad m(0) = m_0 \tag{4.82}$$

$$x(T) = x_T, \quad h(T) = h_T. \tag{4.83}$$

The quantities v_T, γ_T, m_T are free.

Here T denotes the duration of the flight.

Now we formulate the problem of Mayer: in the class of admissible curves Ω, it is required to find those which transfer the trajectory of the system (4.79) from the initial point (4.82) to the point (4.83) and which minimize the functional (4.80).

We solve the problem using the Pontryagin maximum principle (6,18).

We form the funtion H as

$$H = \Psi_1 v \cos\gamma + \Psi_2 v \sin\gamma + \Psi_5 \left(-g \sin\gamma + \frac{v_E\beta}{m}\cos\varepsilon - \frac{D}{m}\right)$$
$$+ \Psi_4 \left(-\frac{g}{v}\cos\gamma + \frac{v_E\beta}{mv}\sin\varepsilon + \frac{L}{mv}\right) - \Psi_5\beta. \tag{4.84}$$

The components of the co-state vector $\Psi(\Psi_1,\ldots,\Psi_5)$ are determined by the following relations:

$$\dot{\Psi}_1 = 0,$$
$$\dot{\Psi}_2 = \frac{1}{m}\frac{\partial D}{\partial h}\Psi_3 - \frac{1}{mv}\frac{\partial L}{\partial h},$$
$$\dot{\Psi}_3 = -\Psi_1 \cos\gamma - \Psi_2 \sin\gamma + \frac{1}{m}\frac{\partial D}{\partial v}\Psi_3$$
$$+ \frac{\Psi_4}{v^2}\left(-g\cos\gamma + \frac{v_E\beta}{m}\sin\varepsilon + \frac{L}{m}\right) - \frac{1}{mv}\frac{\partial L}{\partial v}\Psi_4, \tag{4.85}$$
$$\dot{\Psi}_4 = \Psi_1 v \sin\gamma - \Psi_2 v\cos\gamma + \Psi_3 g \cos\gamma - \Psi_4 \frac{g}{v}\sin\gamma,$$
$$\dot{\Psi}_5 = \frac{\Psi_3}{m^2}(v_E\beta\cos\varepsilon - D) + \frac{\Psi_4}{m^2 v}(v_E\beta\sin\varepsilon + L).$$

The controls $\beta(t)$ and $\varepsilon(t)$ minimizing the fuel expenditure (4.8), must be determined from the condition that the function (4.84) is maximized relative to β and ε.

The condition that H be maximized relative to β determines $\beta(t)$ in the form

$$\beta^0(t) = \begin{cases} \bar{\beta} \text{ for } \Psi_3 \dfrac{v_E}{m}\cos\varepsilon + \Psi_4 \dfrac{v_E}{mv}\sin\varepsilon - \Psi_5 > 0, \\ 0 \text{ for } \Psi_3 \dfrac{v_E}{m}\cos\varepsilon + \Psi_4 \dfrac{v_E}{mv}\sin\varepsilon - \Psi_5 < 0. \end{cases} \tag{4.86}$$

We could have an interval $[t_1, t_2] \in T$ for which

$$\Psi_3 \frac{v_E}{m} \cos \varepsilon + \Psi_4 \frac{v_E}{mv} \sin \varepsilon - \Psi_5 \equiv 0.$$

In this case the control takes on values from the interval (4.81) and is called a singular control. Such controls may be found with the help of the Poisson bracket (15,18).

The condition that H be maximized in ε is equivalent to the relations

$$\frac{\partial H}{\partial \varepsilon} = -\Psi_3 \frac{v_E}{m} \beta \sin \varepsilon + \Psi_4 \frac{v_E}{mv} \beta \cos \varepsilon = 0, \tag{4.87}$$

$$\frac{\partial^2 H}{\partial \varepsilon^2} = -\Psi_3 \frac{v_E}{m} \beta \cos \varepsilon - \Psi_4 \frac{v_E}{mv} \beta \sin \varepsilon < 0. \tag{4.88}$$

The expression (4.87) gives

$$\varepsilon(t) = \operatorname{arctg} \frac{\Psi_4}{\Psi_3 v}, \tag{4.89}$$

while from (4.88) we obtain the inequality

$$\Psi_3 \cos \varepsilon + \Psi_4 \frac{1}{v} \sin \varepsilon > 0, \tag{4.90}$$

which, as is easily seen from (4.89) is satisfied if we take

$$0 \leq \varepsilon \leq 90^\circ.$$

Having determined (4.86) and (4.89), we pass to the transversality conditions (18). In general form, they are written as

$$\left[-\delta m - H\delta t + \Psi_1 \delta x + \Psi_2 \delta h + \Psi_3 \delta v + \Psi_4 \delta \gamma + \Psi_5 \delta m\right]_0^T = 0. \tag{4.91}$$

Applying the boundary conditions (4.82), (4.83), from (4.91) we obtain

$$\Psi_3(T) = 0, \quad \Psi_4(T) = 0, \quad \Psi_5(T) = 1$$
$$H(T) = \Psi_1 v(T) \cos\gamma(T) + \Psi_2(T) v(T) \sin\gamma(T) - \beta(T) = 0. \tag{4.92}$$

For determination of the final form of the laws (4.86), (4.89), it remains to determine the functions $\Psi_3(t)$, $\Psi_4(t)$, $\Psi_5(t)$. This may be done only by integrating the differential equations (4.79), (4.85) using the boundary conditions (4.82), (4.83), (4.92) and keeping in mind the expressions (4.86) and (4.89). For integration of the 10th order system of differential equations (4.79), (4.85), we have 11 boundary conditions (4.82), (4.83), (4.92), defining the constants of integration and the unknown time of the flight T.

b. Flight with Maximal Terminal Velocity. Now we consider the same problem from another point of view. We determine the flight program using another requirement on the flight performance. As an optimality criterion we choose the terminal velocity of the rocket v(T).

In the class of admissible curves Ω, we take those $\beta(t)$ and $\varepsilon(t)$ for which the trajectory of the system (4.79) from the initial point (4.82) comes to the terminal location (4.83) with minimal value of the functional

$$\Delta G = G(t)\Big|_0^T = -(v(T)-v_0). \tag{4.93}$$

Since in this case the expression for the function H in (4.84) remains unchanged, the Eq. (4.85) will be as before and the form of the control laws (4.86) and (4.89) will also not change. Only the condition (4.92) obtained from the transversality conditions will be modified. In place of (4.92), we will have

$$\Psi_3(T) = 1, \quad \Psi_4(T) = 0, \quad \Psi_5(T) = 0,$$
$$H(T) = \Psi_1(T)v(T) \cos \gamma(T) + \Psi_2(T)v(T) \sin \gamma(T) - g \sin \gamma(T)$$
$$+ v_E \frac{\beta(T)}{m(T)} \cos \varepsilon(T) - \frac{D(T)}{m(T)} = 0. \tag{4.94}$$

Thus, the control law for maximizing the terminal velocity will have the form (4.86), (4.89), with the functions $\Psi_3(t)$, $\Psi_4(t)$, $\Psi_5(t)$ coming from integration of the system (4.79), (4.85), under the conditions (4.82), (4.83), (4.94).

c. Minimal-Time Flight. Under only a minimal-time flight requirement, the control law is written in the same form as above (4.86), (4.89). Only the transversality conditions change, now becoming

$$\Psi_3(T) = 0, \quad \Psi_4(T) = 0, \quad \Psi_5(T) = 0,$$
$$H(T) = \Psi_1(T)v(T) \cos \gamma(T) + \Psi_2(T)v(T) \sin \gamma(T) = 1 \tag{4.95}$$

and the solution of the system (4.79), (4.85) is constructed using the boundary conditions (4.82), (4.83), (4.95).

d. Optimal Flight with Respect to Several Criteria. In the preceding point we have considered the problem of rocket flight to a given location optimizing only one criterion chosen in advance. Now we consider the problem of determining a flight program for simultaneous optimization of all the above-considered performance indices, i.e. as functionals for simultaneous optimization we choose

$$I_1(\beta,\varepsilon) = m(T),$$

$$I_2(\beta,\varepsilon) = v(T), \tag{4.96}$$

$$I_3(\beta,\varepsilon) = \int_{t_0}^{T} dt = T.$$

Let each separate functional from (4.96) be optimized according to the previous point and determine the optimal controls β_1^0, ε_1^0 in the problem of maximizing the functional I_1; β_2^0, ε_2^0 for maximizing I_2; β_3^0, ε_3^0 for minimizing I_3. Let these laws lead to the optimal functional values $I_1(\beta_1^0,\varepsilon_1^0) = I_1^0$; $I_2(\beta_2^0,\varepsilon_2^0) = I_2^0$; $I_3(\beta_3^0,\varepsilon_3^0) = I_3^0$, in the space $\{I_1,I_2,I_3\}$, respectively.

According to section 2 of Chapter II, this can be achieved by minimizing the functional

$$R(\beta,\varepsilon) = \left(\frac{I_1(\beta,\varepsilon)-I_1^0}{I_1^0}\right)^2 + \left(\frac{I_2(\beta,\varepsilon)-I_2^0}{I_2^0}\right)^2 + \left(\frac{I_3(\beta,\varepsilon)-I_3^0}{I_3^0}\right)^2 \tag{4.97}$$

over the solutions of the system (4.79) and controls from the region Ω. In this formula I_1^0, I_2^0, I_3^0 are known numbers.

Using the control law determined from minimization of the expression (4.97), some degradation of the values of the performance indices (4.96) from the values $\left\{I_1^0,\ I_2^0,\ I_3^0\right\}$ occurs; however, this deterioration will be minimal and turns out to be uniform in all the considered criteria.

For such a problem statement the Mayer functional is written in the form

$$\Delta G = \left[\left(\frac{m(t)-I_1^0}{I_1^0}\right)^2 + \left(\frac{v(t)-I_2^0}{I_2^0}\right)^2 + \left(\frac{y(t)-I_3^0}{I_3^0}\right)^2\right]_0^T, \tag{4.98}$$

in which $y(t)$ denotes the integral with variable upper limit

$$y(t) = \int_{t_0}^{t} dt.$$

The variable (y) may be considered as the solution of the differential equation

$$\dot{y}(t) = 1 \tag{4.99}$$

with the initial condition

$$y(t_0) = 0. \tag{4.100}$$

The Mayer problem is now the following: from the class of admissible controls Ω, choose $\beta^*(t)$ and $\varepsilon^*(t)$ for which the trajectory of the system (4.79), (4.99), passing from the initial state (4.82) to the final state (4.83), minimizes the functional (4.98).

n this case the function H will have the form

$$H = \Psi_1 v \cos\gamma + \Psi_2 v \sin\gamma + \Psi_3 \left(-g \sin\gamma + \frac{v_E \beta}{m} \cos\varepsilon - \frac{D}{m}\right)$$
$$+ \Psi_4 \left(-\frac{g}{v} \cos\gamma + \frac{v_E \beta}{mv} \sin\varepsilon + \frac{L}{mv}\right) - \Psi_5 \beta + \Psi_6 \tag{4.101}$$

and for determination of the co-state functions $\Psi_1, \ldots, \Psi_6$, we must add the equation

$$\dot{\Psi} = 0. \tag{.102}$$

to the equations (4.85).

It is not difficult to see that in this case the general form of the control law will coincide with (4.86) and (4.89), but for determination of the values of the co-state functions $\Psi_1(t), \ldots, \Psi_6(t)$ we must now solve the system of 12 equations (4.79), (4.85), (4.99), (4.102) under the boundary conditions (4.82), (4.83), (4.100) and the transversality relations

$$2\left(\frac{m(T)-I_1^0}{I_1^0}\right)+\Psi_5(T) = 0,$$

$$2\left(\frac{v(T)-I_2^0}{I_2^0}\right)+\Psi_3(T) = 0,$$

$$2\left(\frac{y(T)-I_3^0}{I_3^0}\right)+\Psi_6(T) = 0, \tag{4.103}$$

$$\begin{aligned} H(T) = {} & \Psi_1(T)v(T)\cos\gamma(T) + \Psi_2(T)v(T)\sin\gamma(T) \\ & + \Psi_3(T)\left(-g\sin\gamma(T) + \frac{v_E\beta(T)}{m(T)}\cos\varepsilon(T) - \frac{D(T)}{m(T)}\right) \\ & - \Psi_5(T)\beta(T) + \Psi_6(T) = 0. \end{aligned}$$

Now we consider a concrete example.

5. Vertical Flight of a Rocket in a Vacuum

As an illustrative example we consider the particular case when $\gamma(T) = \text{const} = 90^0$. In this case, we have $x(t) = \text{const} = x_0$, $\varepsilon(t) = \text{const} = 0$. Since we consider the rocket in a vacuum there is no drag and we will have $D = 0$.

We consider the rocket (154) with parameters $\theta = \dfrac{m_s}{m_s + m_p} = 0.1$ (m_s is the mass of the rocket body, m_p is the mass of fuel), with the effective discharge velocity $v_E = 10^4$ m/sec and with the maximal fuel expenditure rate being $\bar{\beta} = 200$ kg/sec.

The vertical lift of such a rocket in a vacuum is described by the following differential equations

$$\dot{h} = v,$$

$$\dot{v} = 10^4 \frac{\beta}{m} - 10, \qquad (4.104)$$

$$\dot{m} = -\beta,$$

where

$$0 \leq \beta \leq 200. \qquad (4.105)$$

We assume that the gravitational acceleration is constant and equals 10 m/sec^2.

Let the parameters characterizing the initial state of the rocket at time $t_0 = 0$ have the following values:

$$h(0) = h_0 = 0, \quad v(0) = v_0 = 0, \quad m(0) = m_0 = 10^5 kg. \qquad (4.106)$$

We consider the case of rocket flight to an altitude

$$h(T) = h_T = 10^5 m. \qquad (4.107)$$

Here the termination time is free and is to be determined.

At first we consider the problem for the optimization functional being (4.80). In this case, the function H will have the form

$$H = \Psi_1 v + \Psi_2 \left(10^4 \frac{\beta}{m} - 10\right) - \Psi_3 \beta. \qquad (4.108)$$

Maximizing this function over β gives the optimal control law

$$\beta = \begin{cases} 200 \text{ for } k_\beta \geq 0 \\ 0 \text{ for } k_\beta \leq 0. \end{cases} \qquad (4.109)$$

Here k_β denotes $k_\beta = 10^4 \frac{\Psi_2}{m} - \Psi_3$. It is not difficult to see that there are no singular controls for this case.

The equations for determining the co-state functions are

$$\dot{\Psi}_1 = 0,$$

$$\dot{\Psi}_2 = -\Psi_1, \tag{4.110}$$

$$\dot{\Psi}_3 = 10^4 \frac{\beta}{m^2} \Psi_2.$$

It is easy to establish that the function k_β admits only one change of sign. Differentiating this function with respect to time gives

$$\dot{k}_\beta = - 10^4 \frac{\Psi_1}{m} = \frac{A}{m(t)},$$

where

$$A = \text{const} = - 10^4 \Psi, \quad (\Psi_1 = \text{const} \neq 0).$$

Hence, it is seen that k_β is monotonic. This establishes that the optimal control law may have only one switching and, moreover, from physical considerations it may be written in the form

$$\beta = \begin{cases} 200 \text{ for } 0 \leq t \leq \tau \\ 0 \text{ for } \tau \leq t \leq T. \end{cases} \tag{4.111}$$

Here τ denotes the time of switching from the value β = 200 kg/sec to the value $\beta = 0$. For this problem the transversality condition has the form

$$\Big[-\delta m - H\delta t + \Psi_1 \delta h + \Psi_2 \delta v + \Psi_3 \delta m\Big]_0^T = 0,$$

which splits into the following relations

$$\Psi_2(T) = 0, \quad \Psi_3(T) = 1,$$
$$H(T) = \Psi_1(T)v(T) - \beta(T) = 0. \tag{4.112}$$

Consequently, for determination of the fuel expenditure program it is necessary to integrate Eqs. (4.104), (4.110) under the boundary conditions (4.106), (4.107), (4.112).

Since the control can have only one switching, it is sufficient to determine only the unknowns τ and T. This may be done simply by integrating only the system (4.104) with the condition (4.100).

In the interval $0 \leq t \leq \tau$, the trajectory of the system (4.104), (4.111), under the initial condition (4.106), is written in the form

$$m(t) = 10^5 - 200t,$$
$$v(t) = -10t-10^4 \ln(1-0.002t), \tag{4.113}$$
$$h(t) = -5t^2 + 10^4 t + 5 \cdot 10^6 (1-0.002t)\ln(1-0.002t).$$

Using a series expansion for the logarithms, these expressions may be written approximately as

$$m(t) = 10^5 - 200t,$$
$$v(t) = 10t + 0.02t^2, \tag{4.114}$$
$$h(t) = -5t^2 + \frac{2}{3} 20^{-2} t^3.$$

Naturally, the accuracy of the expression (4.114) may be increased if we take more terms in the series expansion of the logarithm function.

The values of the variables (4.114) at time τ must be used as the initial conditions for determining the flight trajectory for $\beta = 0$. Consequently, in the interval $\tau \leq t \leq T$, the form of the flight trajectory is given by the following expressions

$$
\begin{aligned}
m(t) &= \text{const} = m(\tau), \\
v(t) &= 20\tau + 0.02\tau^2 - 10t, \\
h(t) &= -10\tau^2 - \frac{4}{3} 10^{-2}\tau^3 + (20\tau+0.02\tau^2)t - 5t^2.
\end{aligned}
\tag{4.115}
$$

The third relation of (4.12) gives

$$v(T) = 0, \tag{4.116}$$

since for $\tau = T$ we will necessarily have $\beta(T) = 0$ and, as a result of the maximum principle, $\Psi_1(T) = \text{const} \neq 0$.

The last two relation of (4.115) and the conditions (4.107), (4.116) determine two algebraic equations for determination of the unknowns τ and T:

$$
\begin{aligned}
&20\tau + 0.02\tau^2 - 10T = 0, \\
&-10\tau^2 - \frac{4}{3} 10^{-2}\tau^3 + (20\tau+0.02\tau^2)T - 5T^2 = 10^5.
\end{aligned}
\tag{4.117}
$$

The positive solution of this system is

$$
\begin{aligned}
\tau &= 89.3 \text{ sec}, \\
T &= T_1 = 194.55 \text{ sec}.
\end{aligned}
\tag{4.118}
$$

Thus, the optimal flight program (relative to fuel expenditure) of the rocket to an altitude of 100km consists of two regimes: running the motors at maximum capacity for 89.3 sec, followed by switching off the motors for the remainder of the flight. In this case, the desired altitude is achieved with zero terminal velocity and the fuel expenditure is 17860 kg, i.e. we have

$$I_1^0(\beta) = m_0 - m_T = 17860. \tag{4.119}$$

Now we consider the flight problem from another point of view: we now wish to interpret optimality as the maximal velocity of the rocket at the point $h(T) = 10^5$m, which determines the maximal energy of the rocket at this point.

From physical considerations, we note that in this case the flight regime throughout the entire interval $0 \leq t \leq T$ must be with maximal use of the rocket engines.

The mathematical calculation leading to the above result is omitted, as it is quite evident. Thus, we will have

$$\beta(t) = \text{const} = 200 \text{ kg/sec} \tag{4.120}$$

and, from integration of Eq. (4.104), we see that the rocket attains the required altitude $h = 10^5$m at the moment $T_2 = 130.52$ sec with maximal velocity

$$I_2^0(\beta) = v(T) = 1645.9 \text{ m/sec}. \tag{4.121}$$

The fuel expenditure for this flight regime is 26104 kg. Naturally, the minimal-time flight to the point $h = 10^5$m will also come about using the program (4.120).

Now we form the problem of simultaneous optimization of the functionals (4.80) and (4.93).

According to the preceding paragraphs such a problem reduces to the minimzation of the expression

$$R(\beta) = \left(\frac{v(T)-1645.9}{1645.9}\right)^2 + \left(\frac{m_0-m_T-17860}{17860}\right)^2. \tag{4.122}$$

The expressions for the function H and the co-state functions Ψ_1, Ψ_2, Ψ_3 for this case will be the same as that in (4.108), and (4.110), respectively.

The optimal control law coincides in form with (4.111) but we will have correspondingly different values of $\tau = \tau^*$ and $T = T^*$.

In this case, Eqs. (4.104), (4.110) can be integrated under the boundary conditions (4.106) and (4.107), applying the new transversality condition

$$\left[2\left(\frac{v(T)-1645.9}{1645.9^2}\right)\delta v + 2\left(\frac{m_0-m(t)-17860}{17860^2}\right)\delta m - H\delta t+\Psi_1\delta h+\Psi_2\delta v+\Psi_3\delta m\right]_0^T = 0,$$

which, with regard for the boundary conditions, gives

$$\begin{aligned} &2\,\frac{v_T - 1645.9}{1645.9^2} + \Psi_2(T) = 0,\\ &2\,\frac{m_0 - m_T - 17860}{17860^2} + \Psi_3(T) = 0,\\ &H(T) = \Psi_1(T)v(T) - 10\Psi_2(T) = 0. \end{aligned} \tag{4.123}$$

We integrate the system (4.104) for

$$\beta^* = \begin{cases} 200 & \text{for } 0 \le t \le \tau^* \\ 0 & \text{for } \tau^* \le t \le T^*. \end{cases} \tag{4.124}$$

The relations (4.114), (4.115) are written as

$$m(T^*) = m(\tau^*) = 10^5 - 200\tau^*, \tag{4.125}$$

$$v(T^*) = 20\tau^* + 0.02\tau^{*2} - 10T^*, \tag{4.126}$$

$$h(T^*) = -10\tau^{*2} - \frac{4}{3}\,10^{-2}\tau^{*3} + (20\tau^*+0.02\tau^{*2})T^*-5T^{*2} = 10. \tag{4.127}$$

The expression (4.127) connects the two unknown quantities τ^* and T^*. This gives the possibility of reduction to the form

$$T^{*2} - (4\tau^*+0.004\tau^{*2})T^* + 2\tau^{*2} + 2\cdot 10^4 + \frac{8}{3}\,10^{-3}\tau^{*3} = 0, \tag{4.128}$$

the solution of which gives an analytic connection between the unknowns τ^* and T^*

$$T^* = 2\tau^* + 0.002\tau^{*2} - \sqrt{(2\tau^* + 0.002\tau^{*2})^2 - 2\tau^{*2} - \frac{8}{3}\,10^{-3}\tau^{*3} - 2\cdot 10^4}\,. \tag{4.129}$$

It is easy to see that for values of τ^* in the interval $\tau \leq \tau^* \leq T_2$, Eq. (4.128) will have real roots, while the second solution is devoid of physical interest since we can easily establish that this T^* will exceed the value T_1. Now it is easier to substitute the expressions (4.125), (4.126), (4.129) into (4.122) and minimize the latter over τ^*, than to solve the co-state equations (4.110) under the conditions (4.123) and, thus, to establish the final form of the optimal control law (4.124). The expressions (4.122), (4.125), (4.126), (4.129) together determine a one-dimensional function of τ^* of the following form:

TABLE 2

Equations of the object and Boundary Conditions	Functional to be Minimized	Form of the Control Law	Value $m_0 - m_T$ kg	Value v_T m/sec	Value T sec
$\dot{h} = v$ $\dot{v} = 10^4 \frac{\beta}{m} - 10$	$I_1(\beta) = m_0 - m_T [kg]$	$\beta^{(1)} = \begin{cases} 200, & 0 \leq t \leq 89.3 \\ 0, & 89.3 \leq t \leq 194.55 \end{cases}$	17860	0	194.55
$\dot{m} = -\beta$	$I_2(\beta) = -v_T [m/sec]$	$\beta^{(2)} = 200, \; 0 \leq t \leq 130.52$	26104	1645.9	130.52
$0 \leq \beta \leq 200$ kg/sec					
$h(0) = h_0 = 0$					
$v(0) = v_0 = 0$ $m(0) = m_0 = 10^5$ kg	$R(\beta) = \left(\frac{v_T}{1645.9} - 1\right)^2 + \left(\frac{m_0 - m_T}{17860} - 1\right)^2$	$\beta^0 = \begin{cases} 200, & 0 \leq t \leq 117 \\ 0, & 117 \leq t \leq 132.27 \end{cases}$	23400	1291.1	132.27
$h(T) = h_T = 10^5$ m					

$$R(\tau^*) = \left(\frac{10\sqrt{(2\tau^* + 0.002\tau^{*2})^2 - 2\tau^{*2} - \frac{8}{3}\,10^{-3}\tau^{*3} - 2\cdot 10^4}}{1645.9} - 1\right)^2 + \left(\frac{200\tau^* - 17860}{17860}\right)^2. \qquad (4.130)$$

In the interval 89.3 sec $= \tau \le \tau^* \le T_2 = 130.52$ sec, the minimal value of the functional (4.130) is

$$\tau^* = 117 \text{ sec}. \qquad (4.131)$$

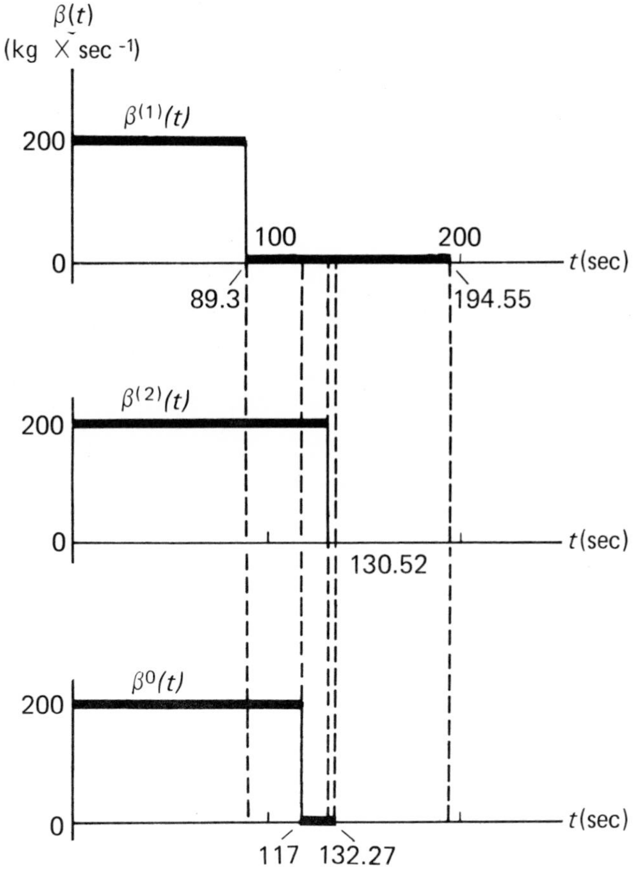

Fig. 10

The final time is now determined from (4.129) as

$$T^* = 132.27 \text{ sec.} \tag{4.132}$$

The terminal velocity of the rocket at $h = 10^5$m is determined from (4.126) and equals $v(T^*) = 1291.1$ m/sec.

For the flight program under the law (4.124), (4.131), (4.132), 23.4 tons of fuel will be used, i.e. we obtain

$$I_1(\beta^*) = m_0 - m_T = 23{,}400 \text{ kg.}$$

For clarity, the results of the calculation are presented in Table 2. Consequently, the flight program defined by the law (4.124), (4.131), (4.132), brings about both a reasonable minimization of fuel expenditure and, at the same time, attainment of a maximal terminal velocity of the rocket. In this case, at the expense of increasing the fuel expenditure by 5540 kg, i.e. by 31% from its limiting minimal value (17860 kg), we attain an increase in the terminal velocity at $h = 10^5$m from 0 to 1291.1 m/sec. On the other hand, decreasing the limiting maximal velocity (1645.9 m/sec) by 354.8 m/sec results in an economy saving of fuel of 2704 kg.

A graphical representation of the control functions $\beta(t)$ for all three flight cases is given in Fig. 10.

CHAPTER V

A. LETOV'S PROBLEM

THE ANALYTIC CONSTRUCTION OF OPTIMAL REGULATORS FOR PROBLEMS WITH VECTOR-VALUED CRITERIA

1. The Problem of Analytic Construction

For the solution of the problem of analytic construction of optimal regulators the basic objective is to obtain a control law by purely analytic means, starting from individual requirements on the transient process, requirements which may be formulated in advance.

As noted in (18), this problem, in general form, is extremely difficult. Therefore, from all the requirements on the transient process, we usually choose only one--the minimization of some measure of the controlled motion.

In general form, such measures are usually described as the sum

$$I(\xi) = \alpha\int_i^f \omega(\eta,\xi,\dot{\xi},t)dt + \beta\Phi(\eta(t_f),\xi(t_f),t_f) \tag{5.1}$$

and we use this sum as a performance index of the system. All the parameters entering into (5.1) have the same meaning as in section 4 of Chapter II.

The functional (5.1) is called an optimality criterion for the regulator and is considered as an axiom determining the content of the theory of analytic construction.

To formulate the requirements on the transient process in the form of several individual general optimality criteria is not easy. The solution obtained by the method of analytic construction, i.e. the form of the control law, depends in form upon the individual optimality criteria and the way they are used to formulate the requirements on the dynamics of the process.

Naturally, for constructing the system engineers must know how to choose the optimality criteria but there are clearly circumstances under which it is impossible to formulate a criterion functional which contains all the basic dynamic properties which the closed system should possess.

In actuality, in the construction of regulators for this or that object, engineers must deal with the following questions: how should one introduce into the controls signals about the coordinates, the velocity, the acceleration? What should be the magnitude of the coefficients of these signals and what should the form of the feedback be? How should we control the regulating mechanism?

In answer to these questions, the builder must start by satisfying those basic performance indices of the regulation process such as the admissible maximal deviation of the regulated quantities, the deviation of the process, the number of oscillations of the regulated quantities and so forth (cf. (2, 155-158)).

All of the above considerations make it necessary to consider the problem of analytic construction for vector-valued criteria, which will promote construction of better regulators. These questions are considered in the works (143,148).

2. The Solution of the General ACOR Problem for Vector Functionals by Dynamic Programming

We consider the perturbed motion of a system described by the differential equations

$$\dot{\eta}_j(t) = f_j(\eta,\xi,t) \quad (j=1,\ldots,n). \tag{5.2}$$

The functions $f_j(\eta,\xi,t)$ $(j=1,\ldots,n)$ are assumed to be continuously differentiable in all their arguments $\eta_1,\ldots,\eta_n$, $\xi_1,\ldots,\xi_m$, t.

Let Eq. (5.2) be defined in an open region for $t \in [0,\infty)$ and let the state variables satisfy the natural boundary conditions

$$\eta_j(0) = \eta_{j0}, \quad \eta_j(\infty) = 0, \quad (j=1,\ldots,n), \quad \sum_{j=1}^{n} \eta_{j0}^2 \le A, \tag{5.3}$$

where A is a given number.

The controlling perturbations lie in the class of piecewise-continuous functions.

Under these conditions, it is required to find a feedback law defined over $t \in [0,\infty]$

$$\xi^0 = \xi^0(\eta_1,\ldots,\eta_n,t),$$

which, along trajectories of the system (5.2), simultaneously minimizes the functionals (2.66).

Finally, we introduce the notation

$$y_\alpha = \int_0^t \omega_\alpha(\eta,\xi)dt \quad (\alpha=1,\ldots,k),$$

so that the problem reduces to minimization of the sum

$$R(\xi) = \sum_{\alpha=1}^{k} \left[y(\infty)-I_{\alpha 0}\right]^2 \tag{5.4}$$

on solutions of the system of n + k equations

$$\begin{aligned} \dot{\eta}_j &= f_j(\eta,\xi,t) \quad (j=1,\ldots,n), \\ \dot{y}_\alpha &= \omega_\alpha(\eta,\xi) \quad (\alpha=1,\ldots,k), \end{aligned} \tag{5.5}$$

under the boundary conditions

$$\eta_j(0) = \eta_{j0}, \quad \eta_j(\infty) = 0, \quad y_\alpha(0) = 0 \qquad (5.6)$$

$$(j=1,\ldots,n;\ \alpha=1,\ldots,k),$$

where η_{j0} $(j=1,\ldots,n)$ are free in (5.3).

In the expression (5.4), as already noted above, the given number $I_{\alpha 0}$ is the value of the functional $I_\alpha(\xi^{(\alpha)})$, obtained as the solution of the ACOR problem for the system (5.2), (5.3) minimizing only over one of the functionals. In other words, I_α is the smallest value of the functional I_α, which can be attained by choosing a control from the class of admissible controls V. The collection of numbers $I_{10}, I_{20}, \ldots, I_{k0}$ forms the utopian point I_0 in the space of functionals.

We introduce the notation

$$2G(y(t) = \sum_{\alpha=1}^{k} \left[y_\alpha(t) - I_{\alpha 0}\right]^2.$$

Then the Mayer functional assumes the following form:

$$\Delta G = G(y,t)\Big|_0^\infty = \frac{1}{2}\sum_{\alpha=1}^{k} (y_\alpha(t) - I_{\alpha 0})^2\Big|_0^\infty . \qquad (5.7)$$

The functional (5.7) is minimized along solutions to the system (5.5) with the boundary conditions (5.6) over controls from the admissible class V.

According to the method of dynamic programming for the problem of Mayer (18), we have the functional equation

$$\min_{\xi \varepsilon V} \left(\frac{\partial \Phi}{\partial t}\right)_{\xi=\xi^0} = 0, \qquad (5.8)$$

in which the function Φ represents

$$\Phi(\eta(\tau),y(\tau),\tau) = \min_{\varepsilon} \Delta G = \min_{\varepsilon} G\Big|_{\tau}^{\infty}$$

and is a function of the variables $\eta_1,\ldots,\eta_n$, $y_1,\ldots,y_k$, t, related by (5.5).

Equation (5.8) is used for determining a control perturbation ξ^0 in the form of a function of $\frac{\partial\Phi}{\partial\eta}$, $\frac{\partial\Phi}{\partial g}$, after which it, together with (5.5), defines the partial differential equation of Hamilton-Jacobi type

$$\frac{\partial\Phi}{\partial t} + \sum_{i=1}^{n} \frac{\partial\Phi}{\partial\eta_i} f_i(\eta,\xi^0,t) + \sum_{\alpha=1}^{k} \frac{\partial\Phi}{\partial y_\alpha} \omega_\alpha(\eta,\xi^0) = 0 \tag{5.9}$$

for the unknown function $\Phi(\eta,y,t)$. At the right end of the trajectory this function must satisfy the condition

$$\Phi(\eta(t),y(t),t)_{t=\infty} = G(y(t))_{t=\infty}. \tag{5.10}$$

We shall find one possible form of the solution of the Hamilton-Jacobi equation. Let $\xi^0 = \xi^0\left(\frac{\partial\Phi}{\partial\eta}\right), \frac{\partial\Phi}{\partial y}$ be a possible form of the optimal control. We substitute it into Eq. (5.9) and seek a solution in the form

$$\Phi(\eta,y,t) = W(\eta,t) + \frac{1}{2}\sum_{\alpha=1}^{k}\left(\frac{\partial G(y(t))}{\partial y_\alpha}\right)_{t=\infty} \frac{\partial G(y(t))}{\partial y_\alpha}. \tag{5.11}$$

We can show that if we can succeed in finding a positive-definite function $W(\eta,t)$ for which the foregoing $\Phi(\eta,y,t)$ satisfies the Hamilton-Jacobi equation, then the control ξ^0 is optimal. In fact, substituting (5.11) into (5.9) gives

$$\frac{\partial W}{\partial t} + \sum_{\alpha=1}^{n} \frac{\partial W}{\partial\eta_\alpha}\dot{\eta}_\alpha = -\sum_{\alpha=1}^{k}\left(\frac{\partial G}{\partial y_\alpha}\right)_{t=\infty} \omega_\alpha(\eta,\xi^0).$$

Since it is trivially clear that

$$\left(\frac{\partial G}{\partial y_\alpha}\right)_{t=\infty} = (y_\alpha(\infty) - I_{\alpha 0}) \geq 0 \quad (\alpha=1,\ldots,k),$$

then

$$\sum_{\alpha=1}^{k} \left(\frac{\partial G}{\partial y_\alpha}\right)_{t=\infty} \omega_\alpha(\eta,\xi^0) \geq 0$$

and, consequently, the function $W(\eta,t)$ will be a Lyapunov function for the closed system.

The question as to whether or not this set-up satisfies the conditions $\eta_j(\infty) = 0$ $(j=1,\ldots,n)$ is not immediately clear. However, in the case when the function W does not depend explicitly on time and satisfies the Barbashin-Krasovskii theorem (159), the stated conditions are satisfied and the ξ^0 constructed above is optimal. When the Lyapunov function W explicitly depends upon t, then we add the supplemental condition $\left(\frac{\partial W}{\partial t}\right)_{t=\infty} = 0$.

Now we show that for determination of the function $\Phi(\eta,y,t)$ in the form (5.11), Eqs. (5.5), (5.8) and conditions (5.10), (5.6) are identical to the equations and conditions obtained earlier for solution of the stated problem by the maximum principle.

According to the maximum principle, the ACOR problem posed in this paragraph reduces to the solution of the system of differential equations

$$\dot{\eta}_j = f_j(\eta,\xi^0,t) \quad (j=1,\ldots,n), \tag{5.12}$$

$$\dot{y}_\alpha = \omega_\alpha(\eta,\xi^0) \quad (\alpha=1,\ldots,k), \tag{5.13}$$

$$\dot{\Psi}_i = -\sum_{j=1}^{n} \Psi_j \frac{\partial f_j}{\partial \eta_i} - \sum_{\alpha=1}^{k} \Psi_{n+\alpha} \frac{\partial \omega_\alpha}{\partial \eta_i} \quad (i=1,\ldots,n), \tag{5.14}$$

$$\dot{\Psi}_{n+\alpha} = 0 \quad (\alpha=1,\ldots,k) \tag{5.15}$$

under the boundary conditions (5.6) and the relations

$$\left[\sum_{i=1}^{n} \Psi_i f_i(\eta, \xi^0, t) + \sum_{\alpha=1}^{k} \Psi_{n+\alpha}\omega_\alpha(\eta, \xi^0)\right]_{t=\infty} = 0, \tag{5.16}$$

$$y_\alpha(\infty) - I_{\alpha 0} + \Psi_{n+\alpha}(\infty) = 0 \quad (\alpha=1,\ldots,k), \tag{5.17}$$

where the function $\xi^0(\eta,t)$ is determined from

$$\max_{\xi\varepsilon V}\left[\sum_{i=1}^{n} \Psi_i f_i(\eta, \xi, t) + \sum_{\alpha=1}^{k} \Psi_{n+\alpha}\omega_\alpha(\eta, \xi)\right] = 0. \tag{5.18}$$

We introduce the notation

$$\Psi_i = -2\frac{\partial\Phi}{\partial\eta_i}, \quad \Psi_{n+\alpha} = -2\frac{\partial\Phi}{\partial y_\alpha} \quad (i=1,\ldots,n;\ \alpha=1,\ldots,k). \tag{5.19}$$

The Eqs. (5.5) coincide with Eqs. (5.12), (5.13). The conditions (5.6) are general. The second relation of (5.19) under (5.11) gives

$$\Psi_{n+\alpha} = -2\frac{\partial\Phi}{\partial y_\alpha} = -\left[y_\alpha(\infty) - I_{\alpha 0}\right] = \text{const} \quad (\alpha=1,\ldots,k),$$

which coincides with conditions (5.17) and determines Eq. (5.15).

In the notation (5.19), identity (5.8) in expanded form is written as

$$\min_{\xi\varepsilon V}\left(\frac{\partial\Phi}{\partial t} - \frac{1}{2}\sum_{i=1}^{n}\Psi_i\dot{\eta}_i - \frac{1}{2}\sum_{\alpha=1}^{k}\Psi_{n+\alpha}\dot{y}_\alpha\right) = 0,$$

which is equivalent to (5.18) since $\frac{\partial\Phi}{\partial t}$ does not explicitly contain t.

From Eq. (5.8) there also arises condition (5.16), since according to the assumptions we have

$$\left.\frac{\partial\Phi}{\partial t}\right|_{t=\infty} = \left.\frac{\partial W}{\partial t}\right|_{t=\infty} = 0. \tag{5.20}$$

It remains to obtain Eq. (5.14). For this, we let the left-side of (5.9) be denoted as z. Clearly, z is a function of $\eta_1,\ldots,\eta_n$, $y_1,\ldots,y_k$, t:

$$z = z(\eta_1,\ldots,\eta_n, y_1,\ldots,y_k, t).$$

Hence, the total differential of z will be

$$dz = \sum_{i=1}^{n} \frac{\partial z}{\partial \eta_i} d\eta_i + \sum_{\alpha=1}^{k} \frac{\partial z}{\partial y_\alpha} dy_\alpha + \frac{\partial z}{\partial t} dt = 0. \tag{5.21}$$

Since Eq. (5.5) determines a system of independent solutions η_i, y_α, (5.21) will occur if and only if the following relations are satisfied

$$\frac{\partial z}{\partial \eta_i} = 0, \quad \frac{\partial z}{\partial y_\alpha} = 0, \quad \frac{\partial z}{\partial t} = 0 \quad (i=1,\ldots,n;\ \alpha=1,\ldots,k). \tag{5.22}$$

Using the first expression in (5.22) for calculating (5.11), (5.19), we obtain

$$\frac{\partial z}{\partial \eta_i} = \frac{\partial}{\partial \eta_i}\left(\frac{\partial \Phi}{\Phi t} + \sum_{j=1}^{n} \frac{\partial \Phi}{\partial \eta_j}\dot{\eta}_j + \sum_{\alpha=1}^{k} \frac{\partial \Phi}{\partial y_\alpha}\dot{y}_\alpha\right)$$

$$= \frac{\partial^2 \Phi}{\partial \eta_i \partial t} - \frac{1}{2}\sum_{j=1}^{n} \frac{\partial \Psi_j}{\partial \eta_i}\dot{\eta}_j - \frac{1}{2}\sum_{j=1}^{n} \Psi_j \frac{\partial \dot{\eta}_j}{\partial \eta_i} - \frac{1}{2}$$

$$\times \sum_{\alpha=1}^{k} \Psi_{n+\alpha} \frac{\partial \dot{y}_\alpha}{\partial \eta_i} = 0 \quad (i=1,\ldots,n),$$

or, what is the same thing

$$\sum_{j=1}^{n} \frac{\partial \Psi_j}{\partial \eta_i} \dot{\eta}_j - 2 \frac{\partial^2 \Phi}{\partial \eta_i \partial t} = - \sum_{j=1}^{n} \Psi_j \frac{\partial \dot{\eta}_j}{\partial \eta_i}$$

$$- \sum_{\alpha=1}^{k} \Psi_{n+\alpha} \frac{\partial \dot{y}_\alpha}{\partial \eta_i} \qquad (i=1,\ldots,n).$$

On the other hand, according to (5.19) and (5.11), we have

$$\dot{\Psi}_i = - \frac{d}{dt}\left(2 \frac{\partial \Phi}{\partial \eta_i}\right) = - 2 \frac{\partial^2 \Phi}{\partial \eta_i \partial t} - 2 \sum_{j=1}^{n} \frac{\partial}{\partial \eta_j}\left(\frac{\partial \Phi}{\partial \eta_i}\right) \dot{\eta}_j$$

$$= - 2 \frac{\partial^2 \Phi}{\partial \eta_i \partial t} + \sum_{j=1}^{n} \frac{\partial \Psi_j}{\partial \eta_i} \eta_j \qquad (i=1,\ldots,n).$$

The last two relations and the defining equations

$$\dot{\Psi}_i = - \sum_{j=1}^{n} \Psi_j \frac{\partial \dot{\eta}_j}{\partial \eta_i} - \sum_{\alpha=1}^{k} \Psi_{n+\alpha} \frac{\partial \dot{y}_\alpha}{\partial \eta_i} \qquad (i=1,\ldots,n),$$

completely coincide with Eqs. (5.14).

Thus, we have shown that for the solution of the ACOR problem for vector functionals by the dynamic programming method, the Bellman function $\Phi(\eta,\xi,t)$ should be sought in the form (5.11) as the solution of the partial differential equation (5.9) under the conditions (5.10), (5.20). For this, the function $W(\eta,t)$ must be positive-definite. It will satisfy all the requirements for a Lyapunov function.

It is of interest to consider linear systems under optimization of vector criteria with components which are quadratic functionals. This question will be studied in a separate paragraph below.

We present a simple example illustrating the items presented above.

3. An Example

We consider a controlled object whose motion is described by the scalar equation

$$\dot{\eta} = \eta + \xi. \tag{5.23}$$

We attempt to synthesize a feedback control $\xi(\eta,t)$ which, on the one hand, yields smooth closed-loop dynamics which are assumed to be solutions of the ACOR problem under the functional

$$I_1(\xi) = \int_0^{\infty} (3\eta^2 + \xi^2)\,dt, \tag{5.24}$$

and, on the other hand, gives the transient process a minimal time characterized by the functional

$$I_2(\xi) = \int_0^{T} dt = T. \tag{5.25}$$

We consider the problem of analytic construction composed with the problem of programming a trajectory.

We will optimize the system with respect to each functional separately. The control minimizing $I_1(\xi)$ will be chosen from the class of continuous functions.

Using the known methods from (18), we find

$$\xi^{(1)} = -3\eta.$$

The control minimizing $I_2(\xi)$ will be taken from the class of piece-wise-continuous, bounded functions taking values in $|\xi| \leq \bar{\xi}$, where $\bar{\xi}$ is a given number.

Application of the maximum principle gives

$$\xi^{(2)} = -\bar{\xi}.$$

Under this control, the system dynamics are

$$\eta = (\eta_0 - \bar{\xi})e^t + \bar{\xi}.$$

Clearly, the time T, corresponding to $\eta(T)$ being zero, exists if and only if $0 \leq \eta_0 \leq \bar{\xi}$.

We consider the process under the above condition. We have

$$T = \ln \frac{\bar{\xi}}{\bar{\xi} - \eta_0} = \ln \frac{1}{1 - \frac{\eta_0}{\bar{\xi}}}.$$

For $\eta_0 < 0$, we compute $\xi^{(2)} = +\bar{\xi}$ and the solution makes sense for $0 < -\eta_0 < \bar{\xi}$. In this case

$$T = \ln \frac{\bar{\xi}}{\bar{\xi} + \eta_0} = \ln \frac{1}{1 + \frac{\eta_0}{\bar{\xi}}}.$$

We consider some cases. Let

$$\eta_0 = \eta(0) = 1, \qquad \bar{\xi} = \max|\xi^{(1)}| = 3.$$

It is not difficult to compute that in this case

$$I_{10} = I_1(\xi^{(1)}) = 3, \qquad I_1^{(2)} = I_1(\xi^{(2)}) \approx 4.087,$$

$$I_{20} = I_2(\xi^{(2)}) = \ln 1.5 \approx 0.405, \qquad I_2^{(1)} = I^2(\xi^{(1)}) = \infty.$$

Now we formulate the problem of optimization of a vector functional. We have the system dynamics

$$\dot{\eta} = \eta + \xi, \quad \dot{y}_1 = 3\eta^2 + \xi^2, \quad \dot{y}_2 = 1, \tag{5.26}$$

where

$$y_1 = \int_0^t (3\eta^2 + \xi^2)dt, \quad y_2 = \int_0^t dt, \quad t \in [0,T]$$

and the boundary conditions

$$\dot{\eta}(0) = 1, \quad \eta(T) = 0, \quad y_1(0) = 0, \quad y_2(0) = 0. \tag{5.27}$$

The Mayer functional for the formulated problem has the following form

$$\Delta G = G(y)\Big|_0^T = \frac{1}{2}\left|(y_1(t) - 3)^2 + (y_2(t) - 0.405)^2\right|_0^T. \tag{5.28}$$

The problem is formulated as: in the class of continuous functions satisfying $|\xi| \leq \bar{\xi} = 3$, find those functions $\xi^0 = \xi^0(\eta,t)$ for which the functional (5.28) assumes its minimum.

In accordance with the dynamic programming method (18), we have

$$\min_{\varepsilon}\left[\frac{\partial\Phi}{\partial t} + (\eta+\xi)\frac{\partial\Phi}{\partial\eta} + (3\eta^2+\xi^2)\frac{\partial\Phi}{\partial y_1} + \frac{\partial\Phi}{\partial y_2}\right] = 0, \tag{5.29}$$

where the function $\Phi(\eta,y_1,y_2,t)$ must be positive-definite and satisfy the boundary condition

$$\Phi(\eta(T),y_1(T),y_2(T),T) = G(y_1(T),y_2(T)). \tag{5.30}$$

Equation (5.29) defines an expression for the optimal control law as

$$\xi^0 = \frac{1}{2}\left(\frac{\partial\Phi}{\partial y_1}\right)^{-1}\frac{\partial\Phi}{\partial\eta}. \tag{5.31}$$

The partial differential equation for the function $\Phi(\eta,y_1,y_2,t)$ has the form

$$\frac{\partial \Phi}{\partial t} + \frac{\partial \Phi}{\partial \eta} \eta - \frac{1}{4} \left(\frac{\partial \Phi}{\partial y_1} \right)^{-1} \left(\frac{\partial \Phi}{\partial \eta} \right)^2 + 3\eta^2 \frac{\partial \Phi}{\partial y_1} + \frac{\partial \Phi}{\partial y_2} = 0. \tag{5.32}$$

The expressions (5.32), (5.31) make sense only if

$$\frac{\partial \Phi}{\partial y_1} \not\equiv 0.$$

Thus, the problem is reduced to the solution of the nonlinear partial differential equation (5.32) under the boundary condition (5.30).

We will seek the function Φ in the form

$$\Phi = \rho_1 (A\eta^2 + f_1(t)\eta + f_2(t)) + \rho_1 (y_1 - 3) + \rho_2 (y_2 - 0.405), \tag{5.33}$$

where ρ_1 and ρ_2 denote

$$\begin{aligned} \rho_1 &= \frac{\partial \Phi}{\partial y_1} = \frac{1}{2} (y_1(T) - 3) > 0, \\ \rho_2 &= \frac{\partial \Phi}{\partial y_2} = \frac{1}{2} (y_2(T) - 0.405) > 0, \end{aligned} \tag{5.34}$$

while the unknowns A, $f_1(t)$, $f_2(t)$ satisfy the conditions

$$A > 0, \quad f_1(t) > 0, \quad t \in [0,T], \quad f_2(T) = \dot{f}_2(T) = 0.$$

These conditions insure asymptotic stability of the closed-loop system (5.23), (5.31) in $[0,\infty]$ and fulfill conditions (5.20), (5.30). It is necessary to note that in the expression (5.32) only the functions $f_1(t)$, $f_2(t)$ explicitly depend on t.

Substituting (5.33) into (5.32), we obtain the relation

$$\rho_1 (2A - A^2 + 3)\eta^2 + \rho_1 (\dot{f}_1 + f_1 - Af_1)\eta + \rho_1 \dot{f}_2 - \frac{1}{4} \rho_1 f_1^2 + \rho_2 = 0,$$

which will occur only when

$$A^2 - 2A - 3 = 0, \tag{5.35}$$

$$\dot{f}_1 - (A-1)f_1 = 0, \tag{5.36}$$

$$\dot{f}_2 - \frac{1}{4} f_1^2 + \frac{\rho_2}{\rho_1} = 0. \tag{5.37}$$

Equation (5.35) determines the solution $A = 3$, while Eq. (5.36) determines the function $f_1(t)$ as

$$f_1(t) = f_1(0)e^{2t}. \tag{5.38}$$

The function $f_2(t)$ is determined from Eq. (5.37) under the condition $f_2(T) = 0$.

It remains to find the initial value $f_1(0)$, after which the feedback law (5.31), solving the problem, may be determined as

$$\xi^0 = -3\eta - \frac{1}{2} f_1(0)e^{2t}. \tag{5.39}$$

The solution of the closed-loop system (5.23) and (5.39), under the initial condition $\eta_0 = 1$, has the form

$$\eta = \left(1 + \frac{1}{8} f_1(0)\right)e^{-2t} - \frac{1}{8} f_1(0)e^{2t}, \tag{5.40}$$

which, according to the boundary condition $\eta(T) = 0$, defines the relation

$$f_1(0) = 8e^{-2T}(e^{2T} - e^{-2T})^{-1}. \tag{5.41}$$

Under the value of $f_1(0)$ determined in (5.41), the expressions (5.39) and (5.40) assume the corresponding forms

$$\xi^0 = -3\eta - 4(e^{2T} - e^{-2T})^{-1}e^{-2(T-t)}, \tag{5.42}$$

$$\eta = (e^{2T} - e^{-2T})^{-1}(e^{2(T-t)} - e^{-2(T-t)}). \tag{5.43}$$

Solving the last two equations of the system (5.26) under the initial conditions (5.27) and applying (5.42), (5.43), we obtain

$$y_1(T) = (e^{2T} - e^{-2T})^{-2}(3e^{4T} - e^{-4T} - 2), \tag{5.44}$$

$$y_2(T) = T. \tag{5.45}$$

According to the condition (5.20), from relation (5.37) it is possible to write

$$f_1^2(T) = 4\frac{\rho_2}{\rho_1}. \tag{5.46}$$

Now relations (5.34), (5.38), (5.44) - (5.46) determine an algebraic equation for the unknown quantity T:

$$\begin{aligned}(T - 0.405)(e^{4T} + e^{-4T} - 2)^2 + 48(e^{4T} + e^{-4T} - 2)& \\ - 16(3e^{4T} - e^{4T} - 2) = 0.&\end{aligned} \tag{5.47}$$

The solution of Eq. (5.47) for T finally determines the control action (5.42), while the values of the functionals (5.24), (5.25) assumed under simultaneous minimization in the accepted sense, are determined by the relations (5.44), (5.45). Solving Eq. (5.47), we obtain the value $T = 0.6975$. The corresponding values of the functionals are

$$I_1 = y_1(T) = 3.262, \quad I_2 = y_2(T) = T = 0.697.$$

Finally, we substitute the numerical value of T into the control action (5.42) and the corresponding trajectory (5.43):

$$\xi^0 = -3\eta - 0.265e^{2t}, \tag{5.48}$$

$$\eta = 1.066e^{-2t} - 0.066e^{2t}. \tag{5.49}$$

The law (5.48) has been determined without considering the constraints on ξ^0. Since, as was seen from (5.48) and (5.49), $|\xi^0|_{t=0} > 3$ and $\xi^0(t) < 0$, the control

$$\xi^0 = \begin{cases} -3 & \text{for } -3\eta - 02.65e^{2t} < -3 \\ -3\eta - 0.265e^{2t} & \text{for } \left|-3\eta - 0.265e^{2t}\right| \leq 3 \end{cases} \tag{5.50}$$

will be suboptimal according to (18).

Naturally, under such a control law the values of the functionals will be different. For their determination we adopt the following procedure.

At first we determine the moment t_{sw} for which the control $\xi^0 = -3$ switches to the control (5.48). To accomplish this, from (5.48), (5.49) we can derive the equation

$$3.198e^{-2t_{sw}} + 0.066e^{2t_{sw}} = 3,$$

having the following solution

$$t_{sw\cdot 1} = 0.038, \quad t_{sw\cdot 2} = 1.895.$$

It makes no sense to consider the second equation, since the process ends before the moment $t_{sw\cdot 2}$.

At the switching time $t_{sw} = 0.038$, the state will have the value

$$\eta(t_{sw}) = -2e^{t_{sw}} + 3 = 0.923. \tag{5.51}$$

We determine those time intervals τ during which the state is determined by (5.49) from the initial state given by (5.51). For this, we must solve the equation

$$1.066e^{-2\tau} - 0.066e^{2\tau} = 0.923,$$

which has the positive solution $\tau = 0.035$.

Thus, the state will be given by (5.49) from the point (5.51) for the time $\Delta t = T - \tau = 0.697 - 0.035 = 0.662$.

The total time of the optimal process is determined by the control (5.50), i.e. the value of the functional (5.25) is determined as

$$I_2^0 = T^0 = t_{sw} + \Delta t = 0.038 + 0.662 = 0.70.$$

Corresponding to the control (5.50), the value of the functional (5.24) is determined as

$$I_1^0 = \int_0^{t_{sw}} \left[3(-2e^t + 3)^2 + 3^2\right] dt$$

$$+ \int_{t_{sw}}^{T^0} \left[3(1.066e^{-2t} - 0.066e^{2t})^2 + (3.198e^{-2t} + 0.066e^{2t})^2\right] dt$$

and takes on the value $I_1^0 = 0.445 + 2.791 = 3.236$.

In Table 3, we present the basic results obtained for this example.

The control law (5.50), determined from the conditions of minimization of the vector functional (5.28), differs noticeably from the control laws $\xi^{(1)} = -3\eta$ and $\xi^{(2)} = -3$, obtained by solving the scalar optimization problems. The increase in the value of the integral criterion (5.24) is approximately 7.86% over its minimal value, while the time of the process is increased from 0.405 to 0.7. Deterioration of both indices takes place simultaneously, and they are uniformly receding from their ideal (utopian) values.

TABLE 3

Equations of the Object and Boundary Conditions	Functional to be Minimized	Control Law
$\dot{\eta} = \eta + \xi$ $\lvert\xi\rvert \le 3$ $\eta(0) = 1,\ \eta(T) = 0$	$I_1 = \int_0^\infty (3\eta^2 + \xi^2)\,dt$	$\xi^{(1)} = -3\eta$
$\dot{\eta} = \eta + \xi$ $\lvert\xi\rvert \le 3$ $\eta(0) = 1,\ \eta(T) = 0$	$I_1 = \int_0^T dt = T$	$\xi^{(2)} = -3$
$\dot{\eta} = \eta + \xi$ $\lvert\xi\rvert < \infty$ $\eta(0) = 1,\ \eta(T) = 0$	$I = \left[\int_0^T (3\eta^2 + \xi^2)\,dt - 3\right]^2 + \left[\int_0^T dt - 0.405\right]^2$	$\xi = -3\eta - 0.265e^{2t}$
$\dot{\eta} = \eta + \xi$ $\lvert\xi\rvert \le 3$ $\eta(0) = 1,\ \eta(T) = 0$	$I = \left[\int_0^T (3\eta^2 + \xi^2)\,dt - 3\right]^2 + \left[\int_0^T dt - 0.405\right]^2$	$\xi^0 = \begin{cases} -3, & 0 \le t \le 0.038; \\ -3\eta - 0.265e^{2t} & 0.038 \le t \le 0.70. \end{cases}$

Equation of the State Trajectory	Value of the Functional $I_1 = \int_0^\infty (3\eta^2 + \xi^2)dt$	Value of the Functional $I_2 = \int_0^T dt = T$
$\eta = e^{-2t}$	3	∞
$\eta = -2e^t + 3$	4.087	$\ln 1.5 \approx 0.405$
$\eta = 1.066e^{-2t} - 0.066e^{2t}$	3.262	0.697
$\eta = \begin{cases} -2e^t + 3 \\ 0 \le t \le 0.038; \\ 1.066e^{-2t} - 0.066e^{2t}, \\ 0.038 \le t \le 0.70. \end{cases}$	3.236	0.70

4. Vector Functionals in Linear ACOR Problems

We set out the solution of linear problems of analytic construction under simultaneous minimization of several quadratic error criteria (148).

Let the perturbed motion of the object in a neighborhood of the equilibrium be described by equations of the form

$$\dot{\eta}_i = \sum_{j=1}^{n} b_{ij}\eta_j + m_i\xi \quad (i=1,\ldots,n), \tag{5.52}$$

where the coefficients b_{ij}, m_i are given constants.

It is assumed that they satisfy the condition that the vectors (160)

$$m, Bm, \ldots, B^{n-1}m, \tag{5.53}$$

are linearly independent, where

$$m = m\{m_1, m_2, \ldots, m_n\}, \qquad B = \left|\left|b_{ij}\right|\right|_1^n.$$

To simplify the computations, we consider a scalar control ξ. The case of a vector control $\xi\{\xi_1,\ldots,\xi_r\}$ contains nothing new, in principle.

Eqs. (5.52) are defined in an open region for $t \in [0,\infty)$ and the states $\eta_1,\ldots,\eta_n$ satisfy the boundary conditions (5.3).

The class of admissible perturbations V is the set continuous functions ξ.

Requirements on the dynamics (5.52) are expressed in the general quadratic criteria

$$I_\alpha(\xi) = \int_0^\infty \left(\sum_{i=1}^{n} \sum_{j=1}^{n} a_{ij}^{(\alpha)} \eta_i\eta_j + \xi \sum_{i=1}^{n} a_i^{(\alpha)}\eta_i + c^{(\alpha)}\xi^2 \right) dt \tag{5.54}$$

$(\alpha = 1,\ldots,k)$.

We assume that if only one of the integrand functions is a positive-definite quadratic form, the remainder may be positive quadratic forms. In some cases these requirements may be relaxed. The values of these functionals characterize a weighted integral of quadratic measures of errors, with weights $\alpha_{ij}^{(\alpha)} \geq 0$, $\alpha_i^{(\alpha)} \geq 0$, $c^{(\alpha)} > 0$, accumulated by the system during its duration. The study of such a problem makes direct physical sense. For example, for the second-order system

$$\ddot{\eta} + 2\alpha_1\dot{\eta} + \alpha_2^2\eta = \xi$$

the functional

$$I_1 = \int_0^\infty (\alpha\eta^2 + b\dot{\eta}^2 + \xi^2)\,dt \qquad (a,b = \text{const} > 0)$$

characterizes an integral of quadratic errors of deviation, while the functional

$$I_2 = \int_0^\infty \ddot{\eta}^2\,dt = \int_0^\infty (\xi - 2\alpha_1\dot{\eta} - \alpha_2^2\eta)^2 dt$$

imposes constraints on the variation of acceleration of the system's motion.

Although these functionals have a similar form, the requirements they express about the system's motion cannot be considered consistently.

<u>Problem.</u> From the class of continuous control perturbations $\xi \in V$, describe those feedback laws

$$\xi^* = \xi^*(\eta_1, \eta_2, \ldots, \eta_n), \tag{5.55}$$

which render the solution of (5.52) asymptotically stable and which optimize the vector functional $I(\xi)$ having components (5.54).

According to section 4 of Chapter II, the solution of the stated problem reduces to the minimization of the sum

$$R(\xi) = \sum_{\alpha=1}^{k} \left(\frac{I_\alpha(\xi)}{I_{\alpha 0}} - 1 \right)^2, \tag{5.56}$$

in which the functional $I_\alpha(\xi)$ is defined as in (5.54), while the $I_{\alpha 0}$ are known positive numbers--the coordinates of the utopian point in the space of functionals. The latter are determined in the following manner: for each individual functional $I_\alpha(\xi)$ $(\alpha=1,\ldots,k)$ from (5.54), we solve the problem of analytic construction of the optimal regulator for the system (5.52) by known methods (18,161). In each case this gives a control

$$\xi^{(\alpha)}(\eta_1,\eta_2,\ldots,\eta_n) = \sum_{i=1}^{n} p_i^{(\alpha)} \eta_i, \tag{5.57}$$

optimizing the chosen functional $I_\alpha(\xi)$ and we calculate the numerical value of this functional using such a control, i.e. we will have

$$I_{\alpha 0} = I_\alpha(\xi^{(\alpha)}) \qquad (\alpha=1,\ldots,k). \tag{5.58}$$

Thus, the stated problem is reduced to the determination of a control (5.55) for which the system (5.52) remains asymptotically stable and for which the functional (5.56) assumes its minimal value.

Solution of the Problem. We introduce the new variables

$$\mu_\alpha(t) = \int_0^\infty \left(\sum_{i=1}^{n} \sum_{j=1}^{n} a_{ij}^{(\alpha)} \eta_i \eta_j + \xi \sum_{i=1}^{n} a_i^{(\alpha)} \eta_i + c^{(\alpha)} \xi^2 \; dt \right) \tag{5.59}$$

$(\alpha=1,\ldots,k)$,

which are the solutions of the differential equations

$$\dot{\mu}_\alpha = -\sum_{i=1}^{n}\sum_{j=1}^{n} \alpha_{ij}^{(\alpha)}\eta_i\eta_j - \xi\sum_{i=1}^{n} a_i^{(\alpha)}\eta_i - c^{(\alpha)}\xi^2 \tag{5.60}$$

$(\alpha=1,\ldots,k)$

under the boundary conditions

$$\mu_\alpha(\infty) = 0 \quad (\alpha=1,\ldots,k). \tag{5.61}$$

Now we formulate the problem of Mayer: among the admissible curves $\eta(t)$, $\mu(t)$, $\xi(t)$, find those which minimize the functional

$$\Delta G = G(\mu(t))\Big|_0^\infty = -\sum_{\alpha=1}^{k}\left(\frac{\mu_\alpha(t)}{I_{\alpha 0}} - 1\right)^2\Big|_0^\infty . \tag{5.62}$$

In correspondence with Eq. (5.9), the dynamic programming functional equation is written in the form

$$\min_{\varepsilon}\left[\sum_{i=1}^{n}\frac{\partial\Phi}{\partial\eta_i}\left(\sum_{j=1}^{n} b_{ij}\eta_j + m_i\xi\right)\right.$$
$$\left. - \sum_{\alpha=1}^{k}\frac{\partial\Phi}{\partial\mu_\alpha}\left(\sum_{i=1}^{n}\sum_{j=1}^{n} a_{ij}^{(\alpha)}\eta_i\eta_j + \xi\sum_{i=1}^{n} a_i^{(\alpha)}\eta_i + c^{(\alpha)}\xi^2\right)\right] = 0, \tag{5.63}$$

which, in turn, determines the optimal feedback law

$$\xi^* = \left(2\sum_{\alpha=1}^{k}\frac{\partial\Phi}{\partial\mu_\alpha}c^{(\alpha)}\right)^{-1}$$
$$\times \sum_{i=1}^{n}\left(\frac{\partial\Phi}{\partial\eta_i}m_i - \sum_{\alpha=1}^{k}\frac{\partial\Phi}{\partial\mu_\alpha}a_i^{(\alpha)}\eta_i\right). \tag{5.64}$$

Here the unknown function Φ is determined as a function of the variables $(\eta_1,\ldots,\eta_n, \mu_1,\ldots,\mu_k)$, satisfying the boundary condition

$$\Phi(\eta(t),\mu(t))_{t=\infty} = G(\mu(t))_{t=\infty}, \tag{5.65}$$

and also the condition

$$\sum_{\alpha=1}^{k} \frac{\partial\Phi}{\partial\mu_\alpha} c^{(\alpha)} \neq 0.$$

<u>Determination of the Function $\Phi(\eta,\mu)$.</u> Eq. (5.63), taken together with (5.64), may be satisfied by the function

$$\Phi^* = \sum_{i=1}^{n} \sum_{j=1}^{n} A_{ij}\eta_i\eta_j + \sum_{\alpha=1}^{k} D_\alpha\mu_\alpha + D_0, \tag{5.66}$$

in which: 1) the coefficients $A_{ij} = A_{ji}$, satisfy the Sylvester condition for positive-definiteness of the quadratic form $W(\eta) = \sum_i \sum_j A_{ij}\eta_i\eta_j$; 2) the coefficients $D_\alpha < 0$, are arbitrary negative numbers; 3) the term D_0 is given as $D_0 = -k$.

We show that a function $W(\eta)$ determined in such a manner may be used as a Lyapunov function for the problem. In fact, its total derivative, according to (5.63), (5.66) and (5.60) is

$$\frac{dW}{dt} = \sum_{\alpha=1}^{k} D_\alpha \left(\sum_{i=1}^{n} \sum_{j=1}^{n} a_{ij}^{(\alpha)}\eta_i\eta_j + \xi\sum_{i=1}^{n} a_i^{(\alpha)}\eta_i + c^{(\alpha)}\xi^2 \right).$$

If even one of the expressions in brackets turns out to be a positive-definite function, then $\frac{dW}{dt}$ will be a negative-definite function, if $D_\alpha < 0$ $(\alpha=1,\ldots,k)$.

Thus, if the function Φ is defined as in (5.66), then the closed-loop system (5.52), (5.64) will be asymptotically stable, which insures satisfaction of the condition (5.3).

Consequently, the boundary condition (5.65) is satisfied for the function equation (5.63).

The expressions (5.63), (5.64) and (5.66) for determination of the unknown coefficients A_{ij}, D_α $(i,j=1,\dots,n;\ \alpha=1,\dots,k)$ give the equation

$$\sum_{i=1}^{n}\sum_{j=1}^{n}\left(\sum_{\gamma=1}^{n} A_{\gamma j}b_{\gamma i} - \sum_{\alpha=1}^{k} D_\alpha a_{ij}^{(\alpha)}\right)\eta_i\eta_j$$

$$+\frac{1}{4\sum_{\alpha=1}^{k} D_\alpha c^{(\alpha)}}\left[\sum_{j=1}^{n}\left(\sum_{i=1}^{n} A_{ij}m_i\right)\eta_j\right.$$

$$\left.-\sum_{i=1}^{n}\left(\sum_{\alpha=1}^{k} D_\alpha a_i^{(\alpha)}\right)\eta_i\right]^2 = 0.$$

The last expression, on account of the independence of the variables $\eta_1,\dots,\eta_n$, splits into the algebraic relation of Riccati-type

$$\sum_{i=1}^{n}\sum_{j=1}^{n}\left[\sum_{\gamma=1}^{n} A_{\gamma j}b_{\gamma i} - \sum_{\alpha=1}^{k} D_\alpha a_{ij}^{(\alpha)}\right.$$

$$+\left(4\sum_{\alpha=1}^{k} D_\alpha c^{(\alpha)}\right)^{-1}\left(\sum_{\gamma=1}^{n} m_\gamma A_{\gamma j}\right)\left(\sum_{\beta=1}^{n} A_{\beta i}m_\beta\right)$$

$$+\left(4\sum_{\alpha=1}^{k} D_\alpha c^{(\alpha)}\right)^{-1}\left(\sum_{\alpha=1}^{k} D_\alpha a_i^{(\alpha)}\right)\left(\sum_{\alpha=1}^{k} D_\alpha a_j^{(\alpha)}\right)$$

$$\left.-\left(2\sum_{\alpha=1}^{k} D_\alpha c^{(\alpha)}\right)^{-1}\left(\sum_{\alpha=1}^{k} D_\alpha a_i^{(\alpha)}\right)\left(\sum_{\gamma=1}^{n} A_{\gamma j}m_\gamma\right)\right]$$

$$\times\ \eta_i\eta_j = 0, \qquad (5.67)$$

with the help of which it is possible to determine the Sylvester coefficients A_{ij} as functions of the values D_α, $\alpha = 1,\ldots,k$.

Now, as follows from formulas (5.64) and (5.65), the solution is represented in the form

$$\xi^* = \sum_{i=1}^{n} \left(\frac{\sum_{j=1}^{n} m_j A_{ij} - \sum_{\alpha=1}^{k} D_\alpha a_i^{(\alpha)}}{2 \sum_{\alpha=1}^{k} D_\alpha c^{(\alpha)}} \right) \eta_i, \tag{5.68}$$

in which the A_{ij} are determined as the solution of the system (5.67), while the coefficients $D_1,\ldots,D_k$ are arbitrary negative numbers satisfying the condition

$$\sum_{\alpha=1}^{k} D_\alpha c^{(\alpha)} \neq 0.$$

The arbitrariness of the negative numbers $D_1,\ldots,D_k$ shows that there exists a k-parameter family $\xi(D_1,\ldots,D_k)$ of optimal feedback laws for the vector problem of analytic construction.

Another interpretation of the constants D_α is the following: D_α ($\alpha=1,\ldots,k$) is the scale factor of the function $I_\alpha(\xi)$ from (5.54). In fact, we multiply each function by the arbitrary multiplier $h_\alpha > 0$. This means that in Eq. (5.67) instead of $a_{ij}^{(\alpha)}$, $a_i^{(\alpha)}$, $c^{(\alpha)}$, we substitute $h_\alpha a_{ij}^{(\alpha)}$, $h_\alpha a_i^{(\alpha)}$, $h_\alpha c^{(\alpha)}$, respectively. We choose h_α so that we have the condition $h_\alpha D_\alpha = -1$ ($\alpha=1,\ldots,k$). Then Eq. (5.67) will not depend upon D_α and h_α, and the solution A_{ij} ($i,j,=1,\ldots,n$) and also the control ξ^*, determined in (5.68), will be only functions of the parameters of the system (5.52) and the weighting coefficients of the functionals (5.54).

Therefore, in (5.67) and (5.68) we should have formally set $D_\alpha = -1$ ($\alpha=1,\ldots,k$). However, this is impossible to do for the following reason: in the solution of the problem of minimizing the scalar functional $I_\alpha(\xi)$, there are no principal values which minimize $-I_\alpha(\xi)$ or $h_\alpha I_\alpha(\xi)$, where h_α = const > 0. In the vector optimization problem, there results an approximation of the values of all the functionals (5.54) to their numerical values (5.58), which are obtained by optimizing the separate scalar functionals. Since the original values η_{i0} are not fixed and are constrained only by the condition $\sum_{i=1}^{n} \eta_{i0}^2 \leq A$, the coordinates of the utopian point (5.58) in any individual case may be selected using additional considerations concerning choice of an "average" point in the domain of A.

<u>Another Approach to the Problem.</u> The values of the parameters $D_1,\ldots,D_k$ as scale factors is especially clear in other approaches to the vector optimization problem.

Let there be given the set of scalar functionals (5.54). From them we form a linear combination with weighting coefficients $D_1,\ldots,D_k$.

It is not difficult to show that the solution (5.68) is also obtained in the case of the usual problem of analytic construction for the scalar functional

$$I(\xi) = -\sum_{\alpha=1}^{k} D_\alpha I_\alpha(\xi). \tag{5.69}$$

The last expression represents an integral of quadratic measures of error with weighting coefficients depending on $D_1 < 0,\ldots,$ $D_k < 0$:

$$I(\xi) = \int_0^\infty \left(\sum_{i=1}^{n} \sum_{j=1}^{n} a^*_{ij}\eta_i\eta_j + \xi \sum_{i=1}^{n} a^*_i\eta_i + c^*\xi^2 \right) dt. \tag{5.70}$$

These coefficients a^*_{ij}, a^*_i, c^*_i determine positive integrand functions, are finite and depend on the values $D_1,\ldots,D_k$ since

$$a^*_{ij} = -\sum_{\alpha=1}^{k} D_\alpha a^{(\alpha)}_{ij}, \quad a^*_i = -\sum_{\alpha=1}^{k} D_\alpha a^{(\alpha)}_i, \quad c^* = -\sum_{\alpha=1}^{k} D_\alpha c^{(\alpha)}.$$

In such a problem we will have a Lyapunov function in the form of a quadratic form with unknown coefficients depending on $D_1, \ldots, D_k$,

$$W^*(\eta_1, \ldots, \eta_n) = \sum_{i=1}^{n} \sum_{j=1}^{n} A^*_{ij}(D)\eta_i\eta_j. \tag{5.71}$$

Using this function, we determine the values of the functionals (5.70) as

$$I(\xi^*) = W^*(\eta_{10}, \ldots, \eta_{n0}) = \sum_{i=1}^{n} \sum_{j=1}^{n} A^*_{ij}(D)\eta_{i0}\eta_{j0} \tag{5.72}$$

and the form of the optimal control

$$\xi^* = -\frac{1}{2c^*} \sum_{i=1}^{n} \left(\sum_{i=1}^{n} A^*_{ij}(D)m_j + a^*_i \right) \eta_i. \tag{5.73}$$

We assume that $D_1, \ldots, D_k$, treated as weighting coefficients of the functional (5.69), are known, given numbers. Then the coefficients A^*_{ij} $(i,j=1,\ldots,n)$ in (5.71) - (5.73) may be determined from the Riccati equation (5.67), in which A_{ij} are substituted by A^*_{ij}, respectively. Consequently, the control (5.73) is determined as a function of the parameters $D_1, \ldots, D_k$. Now we calculate the values of the functionals (5.54) along the trajectory of the system (5.52) operating under the control law (5.73). It is not difficult to show that these values $I_\alpha(\xi^*)$ $(\alpha=1,\ldots,k)$ may be chosen as quadratic forms

$$I_\alpha(\xi^*) = \sum_{i=1}^{n} \sum_{j=1}^{n} B^{(\alpha)}_{ij} \eta_{i0}\eta_{j0}, \tag{5.74}$$

the coefficients of which, for each fixed α, are determined by comparing the quadratic form

$$-\sum_{i=1}^{n}\left[\frac{\partial}{\partial\eta_i}\left(\sum_{i=1}^{n}\sum_{j=1}^{n}B_{ij}^{(\alpha)}\eta_i\eta_j\right)\right]\left(\sum_{j=1}^{n}b_{ij}\dot{\eta}_j+m_i\xi^*\right)$$

with the integrand of the functional I_α for $\xi = \xi^*$ in (5.73). Clearly, these coefficients $B_{ij}^{(\alpha)}$ $(i,j=1,\ldots,n;\ \alpha=1,\ldots,k)$ will be determined as functions of the parameters $D_1,\ldots,D_k$.

Finally, having determined (5.74), the functional (5.56) is written in the form

$$R(\xi) = \sum_{\alpha=1}^{k}\left(\frac{\sum_{i=1}^{n}\sum_{j=1}^{n}B_{ij}^{(\alpha)}\eta_{i0}\eta_{j0}}{I_{\alpha 0}} - 1\right)^2 \qquad (5.75)$$

which may be considered as a known function of the parameters $D_1,\ldots,D_k$. From the condition that (5.75) be minimized with respect to D_α, we may determine the unknown parameters $D_1,\ldots,D_k$. These parameters determine the final form of the solution (5.73). We note that the results of the work may be generalized to the case of systems having constantly acting external disturbances (162-165) and to systems with delays (166-168).

5. An Example

For clarity, we consider a control object, the perturbed motion of which is described by the system of equations

$$\dot{\eta}_1 = \eta_2, \quad \dot{\eta}_2 = \eta_1 + \eta_2 + \xi. \qquad (5.76)$$

We try to synthesize a control $\xi^*(\eta_1,\eta_2)$ for which the closed-loop system is asymptotically stable and, at the same time, minimizes the functionals

$$I_1(\xi) = \int_0^\infty (8\eta_1^2 + \xi^2)\,dt, \qquad I_2(\xi) = \int_0^\infty (11\eta_2^2 + \xi^2)\,dt. \tag{5.77}$$

It is not difficult to calculate that a control making (5.76) asymptotically stable and minimizing $I_1(\xi)$ will have the form

$$\xi^{(1)} = -4\eta_1 - 4\eta_2, \tag{5.78}$$

while to minimize only the functional $I_2(\xi)$ we have

$$\xi^{(2)} = -2\eta_1 - 5\eta_2. \tag{5.79}$$

The minimal values of these functionals are determined, respectively, by the quadratic forms

$$\begin{aligned} I_{10}(\xi^{(1)}) &= 8\eta_{10}^2 + 8\eta_{10}\eta_{20} + 4\eta_{20}^2, \\ I_{20}(\xi^{(2)}) &= 3\eta_{10}^2 + 4\eta_{10}\eta_{20} + 5\eta_{20}^2. \end{aligned} \tag{5.80}$$

We form a linear combination of the functionals (5.77)

$$I(\xi) = D_1 I_1(\xi) + D_2 I_2(\xi). \tag{5.81}$$

Under the assumption that $D_1 = D_2 = 1$, the solution of the analytic construction problem for the system (5.76), with functional (5.81), will have the form

$$\xi = -3.24\eta_1 - 4.6\eta_2. \tag{5.82}$$

Now we determine the control $\xi^*(\eta_1,\eta_2)$ optimizing the vector functional with components (5.77). As noted in section 4, the problem reduces to the determination of the scale multipliers D_1, D_2 in the functional (5.81). We write the functional (5.81) in the expanded form

$$I(\xi) = \int_0^\infty \left[8D_1\eta_1^2 + 11D_2\eta_2^2 + (D_1+D_2)\xi^2\right]dt \tag{5.83}$$

and solve the ordinary problem of analytic construction. The functional equation for (5.83) is written in the form

$$\frac{\partial W}{\partial \eta_1}\eta_2 + \frac{\partial W}{\partial \eta_2}(\eta_1 + \eta_2 + \xi) + 8D_1\eta_1^2 + 11D_2\eta_2^2 + (D_1 + D_2)\xi^2 = 0. \tag{5.84}$$

Eq. (5.84) determines the optimal control (5.73) as

$$\xi^* = -\frac{1}{2(D_1 + D_2)}\frac{\partial W^*}{\partial \eta_2}. \tag{5.85}$$

The Lyapunov function (5.71) must be sought in the form of the quadratic form

$$W^*(\eta_1,\eta_2) = A_{11}^*\eta_1^2 + 2A_{12}^*\eta_1\eta_2 + A_{22}^*\eta_2^2. \tag{5.86}$$

For determination of the coefficients, (5.84), (5.85) and (5.86) give the system of algebraic equations

$$\begin{aligned}
&A_{12}^{*2} - 2(D_1 + D_2)A_{12}^* - 8D_1(D_1 + D_2) = 0,\\
&A_{22}^* - 2(D_1 + D_2)A_{22}^* - \left[2(D_1 + D_2)A_{12}^* + 11D_2(D_1 + D_2)\right] = 0,\\
&A_{11}^* - \left(\frac{1}{D_1 + D_2}A_{12}^*A_{22}^* - A_{12}^* - A_{22}^*\right) = 0.
\end{aligned} \tag{5.87}$$

Eq. (5.87) allows us to determine the coefficients of the quadratic form (5.86) in the form of functions of the positive parameters D_1, D_2. It is not difficult to compute that they will be

$$A_{12}^* = D_1 + D_2 + \sqrt{9D_1^2 + 10D_1D_2 + D_2^2},$$

$$A_{22}^* = D_1 + D_2 + \sqrt{3D_1^2 + 17D_1D_2 + 14D_2^2 + 2(D_1+D_2)\sqrt{9D_1^2 + 10D_1D_2 + D_2^2}}, \tag{5.88}$$

$$A_{11}^* = \frac{1}{D_1 + D_2} A_{12}^* A_{22}^* - A_{12}^* - A_{22}^*.$$

We determine the values of the functionals $I_1(\xi^*)$ and $I_2(\xi^*)$, where ξ^* is determined from (5.85), (5.86) and (5.88) as

$$\xi^* = -\frac{A_{12}^*}{D_1 + D_2}\eta_1 - \frac{A_{22}^*}{D_1 + D_2}\eta_2. \tag{5.89}$$

We find that

$$I_1(\xi^*) = B_{11}^{(1)}\eta_{10}^2 + 2B_{12}^{(1)}\eta_{10}\eta_{20} + B_{22}^{(1)}\eta_{20}^2, \tag{5.90}$$

in which

$$B_{11}^{(1)} = \frac{\left[A_{12}^{*2} + 8(D_1 + D_2)^2\right](A_{22}^* - D_1 - D_2)}{2(D_1 + D_2)^2(A_{12}^* - D_1 - D_2)} + \frac{A_{12}^{*2} + 8(D_1 + D_2)^2}{2(D_1 + D_2)(A_{22}^* - D_1 - D_2)} + \frac{A_{22}^*(A_{12}^* - D_1 - D_2)}{2(D_1 + D_2)^2(A_{22}^* - D_1 - D_2)} - \frac{A_{12}^* A_{22}^*}{(D_1 + D_2)^2}, \tag{5.91}$$

$$B_{12}^{(1)} = \frac{A_{12}^{*2} + 8(D_1 + D_2)^2}{2(D_1 + D_2)(A_{12}^* - D_1 - D_2)} ,$$

$$B_{22}^{(1)} = \frac{A_{12}^{*2} + 8(D_1 + D_2)^2}{2(A_{12}^* - D_1 - D_2)(A_{22}^* - D_1 - D_2)} + \frac{A_{22}^{*2}}{2(D_1 + D_2)(A_{22}^* - D_1 - D_2)} .$$

Analogously, for the second function we have

$$I_2(\xi^*) = B_{11}^{(2)} \eta_{10} + 2B_{12}^{(2)} \eta_{10}\eta_{20} + B_{22}^{(2)} \eta_{20}^2, \tag{5.92}$$

where

$$B_{11}^{(2)} = \frac{A_{12}^{*2}\ (A_{22}^* - D_1 - D_2)}{2(D_1 + D_2)^2 (A_{12}^* - D_1 - D_2)} + \frac{A_{12}^{*2}}{2(D_1 + D_2)(A_{22}^* - D_1 - D_2)}$$

$$+ \frac{\left[A_{22}^{*2} + 11(D_1 + D_2)^2\right](A_{12}^* - D_1 - D_2)}{2(D_1 + D_2)^2 (A_{22}^* - D_1 - D_2)} - \frac{A_{12}^* A_{22}^*}{(D_1 + D_2)^2} , \tag{5.93}$$

$$B_{12}^{(2)} = \frac{A_{12}^{*2}}{2(D_1 + D_2)(A_{12}^* - D_1 - D_2)} ,$$

$$B_{22}^{(2)} = \frac{A_{12}^{*2}}{2(A_{12}^* - D_1 - D_2)(A_{22}^* - D_1 - D_2)} + \frac{A_{22}^{*2} + 11(D_1 + D_2)^2}{2(D_1 + D_2)(A_{22}^* - D_1 - D_2)} .$$

According to the expressions (5.80), (5.88), (5.90) - (5.93), the functional (5.75) may be written as the following function of the parameters

$$R(\xi) = R(D_1, D_2) = \left(\frac{B_{11}^{(1)} \eta_{10}^2 + 2B_{12}^{(1)} \eta_{10}\eta_{20} + B_{22}^{(1)} \eta_{20}^2}{8\eta_{10}^2 + 8\eta_{10}\eta_{20} + 4\eta_{20}^2} - 1 \right)^2$$
$$+ \left(\frac{B_{11}^{(2)} \eta_{10}^2 + 2B_{12}^{(2)} \eta_{10}\eta_{20} + B_{22}^{(2)} \eta_{20}^2}{3\eta_{10}^2 + 4\eta_{10}\eta_{20} + 5\eta_{20}^2} - 1 \right)^2 . \tag{5.94}$$

TABLE 4

Control Object	Functional to be Minimized
$\dot{\eta}_1 = \eta_2$ $\dot{\eta}_2 = \eta_1 + \eta_2 + \xi$	$I(\xi) = D_1 \int_0^\infty (8\eta_1^2 + \xi^2)\,dt$ $+ D_2 \int_0^\infty (11\eta_2^2 + \xi^2)\,dt$ $= \left(\dfrac{\int_0^\infty (8\eta_1^2 + \xi^2)\,dt}{8\eta_{10}^2 + 8\eta_{10}\eta_{20} + 4\eta_{20}^2} - 1 \right)^2$ $+ \left(\dfrac{\int_0^\infty (11\ {}_2^2 + {}^2)\,dt}{3\eta_{10}^2 + 4\eta_{10}\eta_{20} + 5\eta_{20}^2} - 1 \right)^2$

The solution of the equations

$$\frac{\partial R(D_1,D_2)}{\partial D_1} = 0, \qquad \frac{\partial R(D_1,D_2)}{\partial D_2} = 0 \tag{5.95}$$

in the domain $0 \le D_1, D_2 \le 1$, under the given expressions (5.88), (5.91), (5.93), (5.94), determines the values of the parameters D_1 and D_2. Indeed, for initial points of the state in the (η_1,η_2)-space given by the equation $\eta_{10} = \eta_{20}$, the numerical solution of Eq. (5.95) is

$$D_1 = 0.472; \quad D_2 = 0.854. \tag{5.96}$$

Domain of Initial Values	Values of the Parameters D_1 & D_2	Form of the Feedback Control
$-\infty < \eta_{10} < \infty$, $-\infty < \eta_{20} < \infty$	$D_1 = 1;\ D_2 = 0$	$\xi = -4\eta_1 - 4\eta_2$
	$D_1 = 0;\ D_2 = 1$	$\xi = -2\eta_1 - 5\eta_2$
	$D_1 = 1;\ D_2 = 1$	$\xi = -3.24\eta_1 - 4.6\eta_2$
$\eta_{10} = \eta_{20}$	$D_1 = 0.472;\ D_2 = 0.854$	$\xi = -2.96\eta_1 - 4.74\eta_2$
$\eta_{10} = 2\eta_{20}$	$D_1 = 0.326;\ D_2 = 0.685$	$\xi = -2.89\eta_1 - 4.77\eta_2$
$\eta_{10} = 5\eta_{20}$	$D_1 = 0.239;\ D_2 = 0.559$	$\xi = -2.84\eta_1 - 4.79\eta_2$
$\eta_{10} = 10\eta_{20}$	$D_1 = 0.236;\ D_2 = 0.569$	$\xi = -2.83\eta_1 - 4.8\eta_2$

These values D_1, D_2 determine the optimal law (5.89) in the form

$$\xi^* = -2.96\eta_1 - 4.74\eta_2. \tag{5.97}$$

Analogously, for initial points given by the equation $\eta_{10} = 2\eta_{20}$, we obtain

$$D_1 = 0.326; \quad D_2 = 0.685; \tag{5.98}$$

$$\xi^* = -2.89\eta_1 - 4.77\eta_2. \tag{5.99}$$

For initial points given by the equation $\eta_{10} = 5\eta_{20}$, we obtain

$$D_1 = 0.239; \quad D_2 = 0.559; \tag{5.100}$$

$$\xi^* = -2.84\eta_1 - 4.79\eta_2. \tag{5.101}$$

For initial points given by $\eta_{10} = 10\eta_{20}$, we have

$$D_1 = 0.236; \quad D_2 = 0.569; \tag{5.102}$$

$$\xi^* = -2.83\eta_1 - 4.80\eta_2. \tag{5.103}$$

The numerical solution of the system of equations (5.95) in the different cases and the determination of the values (5.96), (5.98), (5.100) and (5.102), was carried out using the method presented in (169).

For completeness and clarity, the basic results obtained for the solution of the considered example are presented in Table 4.

As is seen, expressions (5.97), (5.99), (5.101), (5.103) differ substantially from (5.78), (5.79), (5.82).

CHAPTER VI

THE OPTIMIZATION OF VECTOR FUNCTIONALS IN LINEAR (NONLINEAR) PROGRAMMING PROBLEMS

1. The Linear Programming Problem

One of the important areas of optimization is linear programming, which has been successfully applied for the solution of a series of important modern control problems in economics, automated production processes, planning and so on. Linear programming methods were widely developed following the work of L.V. Kantorovich (170,171). Currently, the mathematics is well-understood and there are effective procedures for optimizing scalar objective functions (25,27,172,173). Optimization of vector objective functions is treated in (180-185).

The general linear programming problem is mathematically formulated in the following manner. Let there be given a linear function of n variables $X(x_1,\ldots,x_n)$

$$L(X) = c_1x_1 + c_2x_2 +\ldots+ c_nx_n \tag{6.1}$$

and linear constraints imposed on the variables

$$\alpha_{\alpha 1}x_1 + \alpha_{\alpha 2}x_2 +\ldots+ \alpha_{\alpha n}x_n = b_\alpha \qquad (\alpha=1,\ldots,r),$$

$$\alpha_{r+\beta,1}x_1 + \alpha_{r+\beta,2}x_2 +\ldots+ \alpha_{r+\beta,n}x_n \leq b_{r+\beta} \tag{6.2}$$

$$(\beta=1,2,\ldots,m-r), \tag{6.3}$$

$$x_1 \geq 0, x_2 \geq 0,\ldots,x_s \geq 0 \qquad (s\leq n). \tag{6.4}$$

The matrix $||\alpha_{ij}||$ composed of the coefficients of the system (6.2), (6.3), is assumed to be given and is called the matrix of problem conditions. The linear form (6.1) is called the objective function. The problem consists of the following: among the solutions of the system (6.2) - (6.4), forming a region Ω in the space X, find those for which the form (6.1) is maximized (or minimized). The solution $X^0 \varepsilon \Omega$, is called the optimal plan of the problem.

The linear programming problem serves as a mathematical model of many important control problems arising in practice. It is solved either exactly, or approximately, by Dantzig's simplex method (27,173). In the Russian literature, this is often called the method of successive improvements of the plan.

However, in many practical optimization problems, it is not sufficient to have only one system performance index. More acute is the requirement of optimization of two or more performance indices of the system, which leads to an optimization problem with a vector performance criterion. In the transportation problem, for instance, it is possible to identify two performance indices: the total cost of transferring goods and the time needed for supplying goods to the consumers (173). If an optimal plan is constructed in the first case with the goal of maximal reduction of costs, then in the second the optimal plan is assumed to be that for which all goods may be supplied to demand points in the shortest time. The two performance indices of the transfer operation are in conflict. Nevertheless, it is impossible to avoid situations in which it is desired to obtain a solution of the transfer problem in "good" time satisfying a plan and at sufficiently low cost. The collection of all these requirements may be accounted for by some choice of objective function, which forms a vector performance index--a vector objective function.

2. The General Programming Problem

In (21), we find the following general statement of a programming problem: we consider a vector $X \in E^n$ and a convex set $\Omega \subset E^n$. Assume that we are given functions $g(X)$ and $f_\alpha(X)$, $\alpha = 1,\ldots,m$, where m is a given integer.

The mathematical programming problem is formulated as follows: minimize (or maximize) the function $g(X)$ under the conditions

$$X \in \Omega, \quad f_\alpha(X) \geq 0 \quad (\alpha=1,\ldots,m). \tag{6.5}$$

If $g(X)$, $f_\alpha(X)$ are linear functions as in (6.1) - (6.4), then the given problem reduces to the linear programming problem; if $g(X)$, $f_\alpha(X)$ are arbitrary (convex) functions, then the given problem is called a nonlinear (convex) programming problem.

3. A More General Problem

We assume that a vector function $G(X) = \{g_1(X),\ldots,g_k(X)\}$ is given on the set Ω. We call the vector X admissible if it satisfies the constraints (6.5). We will say that an admissible vector X^0 is an effective point if there does not exist another admissible vector X for which

$$G(X^0) \leq G(X). \tag{6.6}$$

The inequality (6.6) should be understood as

$$g_\alpha(X^0) \leq g_\alpha(X) \quad (\alpha=1,\ldots,k), \tag{6.7}$$

where the inequality is strict for at least one of the α's.

The programming problem for a vector functional consists of searching for all effective points.

In the case when Ω is a closed, convex set, while the functions $f_\alpha(X)$, $g_\beta(X)$ $(\alpha=1,\ldots,m;\ \beta=1,\ldots,k)$ are concave functions of the variables $x_1,\ldots,x_n$, the following lemma is proved in (21): if X^0 is an effective point for the problem (6.5), (6.6), then there exists a vector $v\{v_1,\ldots,v_n\}$ with components $v_i \geq 0$, satisfying the equation $\sum_{\alpha=1}^{n} v_\alpha = 1$, such that the maximum of the scalar function $g(X) = (v,G(X))$ over the set of all X satisfying (6.5) is attained for $X = X^0$. The collection of all effective points obtained for different values of the vector v forms the Pareto-optimal hyperplane in the space G(X). To each point of this hyperplane corresponds an effective point from the space Ω.

In the same work, a theorem is proved permitting the reduction of the vector optimization problem to the search for saddle-points of a specially constructed scalar function.

4. Ordered Sets in Programming Problems

In (102) we find further development of the idea of an effective point, which was studied in the preceding paragraph. As was already noted in section 3 of Chapter I, in the case of a programming problem the idea of an ordered set of functionals is very frequently encountered in operations research problems.

We give a definition of the concept of an ordered set in connection with programming problems. With this objective in mind, we turn to relations (6.5), (6.6). Let Y be an admissible set of values for x. We will say that the set of scalar criteria $g_1(x),\ldots,g_k(x)$ is ordered if there exist subsets

$$Y_k \subset Y_{k-1} \subset \ldots \subset Y_1 \subset Y,$$

on which are assumed

$$\inf g_k(x), \quad x \in Y_k,$$

$$\inf g_{k-1}(x), \quad x \in Y_{k-1},$$

$$\cdots\cdots\cdots\cdots$$

$$\inf g_1(x), \quad x \in Y_1,$$

where each subset Y_k is non-empty.

The interpretation of this definition is exactly as in the continous case of programming optimal trajectories, which was studied in (59).

The basic result in the work (102) is that any solution of the programming problem with respect to the ordered set of criteria $g_1(x),\ldots,g_k(x)$ is an effective point x^0 for the same system of criteria, i.e. condition (6.7) is satisfied. In this same work, a method is developed allowing us to construct a scalar functional

$\overline{G}(x) = \sum_{\alpha=1}^{k} \lambda_\alpha g_\alpha(x)$, with $\lambda_\alpha > 0$, which reduces the optimization problem for the set of criteria $g_1(x),\ldots,g_k(x)$ to the optimization of $\overline{G}(x)$. In connection with this method, we note only that it is computationally very difficult to implement.

5. A Different Statement of the Linear (Nonlinear) Programming Problem for Vector Criteria

We now explain the statement of the linear (nonlinear) programming problem with vector criteria, which was described earlier in (145). At its basis lies the idea of approximation, which was studied in Chapter II of this monograph.

Let there be given the system of linear forms

$$L_i(X) = \sum_{j=1}^{n} c_{ij}x_j \quad (i=1,\ldots,k) \tag{6.8}$$

and the boundary conditions (6.2) - (6.4), defining the domain of values of the variables $x_1,\ldots,x_n$. We note that the forms $L_1(X)$, $\ldots,L_k(X)$ may have different dimensions.

Among the solutions of the system (6.2) - (6.4), it is required to find those values of the vector $X^*\{x_1^*,\ldots,x_k^*\} \in \Omega$, for which the linear forms (6.8) take on their maximal (minimal) values simultaneously.

We consider each individual form of (6.8) as a scalar function and assume that for each fixed i (i=1,...,k), the solution of the standard linear programming problem is given. Let the corresponding optimal plan be characterized by the vectors

$$X_i^0(x_{i1}^0,\ldots,x_{in}^0) \quad (i=1,\ldots,k). \tag{6.9}$$

Using this plan, we determing the values of the linear forms (6.8) as

$$L_1^0 = L_1(X_1^0), L_2^0 = L_2(X_2^0),\ldots,L_k^0 = L_k(x_k^0). \tag{6.10}$$

Thus, for example, the vector $X_\alpha^0(x_{\alpha 1}^0,x_{\alpha 2}^0,\ldots,x_{\alpha n}^0)$ is the solution of the minimization problem for the linear form $L_\alpha(X)$ from (6.8), under the constraints (6.2) - (6.4) and the numerical value of this form is $L_\alpha^0 = L_\alpha(X_\alpha^0)$. Naturally, the vectors (6.9), determining a point in the space of variables $(x_1,\ldots,x_n) \in \Omega$, will be different, although some of them may coincide with others.

We consider the vector L(X) with components $L_i(X)/L_i^0$ from (6.8) and form the euclidean norm

$$R(X) = ||L(X) - L^0||^2 = \sum_{i=1}^{k} \left[\sum_{j=1}^{n} \frac{c_{ij}}{L_i^0} x_j - 1 \right]^2 \tag{6.11}$$

The vector $L(X) - L^0$ is defined for all $X \in \Omega$. We note that L^0 will be a unit vector in the space of vectors L(X). We call it the ideal or utopian value of the vector L(X).

Definition. We will say that the plan $X^*(x_1^*,\ldots,x_n^*)$ optimizes the vector performance index with components (6.8) if

$$R(X^*) \leq R(X) \tag{6.12}$$

for all $X \in \Omega$.

We call such a plan a vector-optimal plan for the linear programming problem.

Now we re-state the problem just formulated as: given the system of linear forms (6.8) and the conditions (6.2) - (6.4), determine points $X^* \in \Omega$, at which the function R(X) is minimal.

Geometrical Interpretation of the Problem. We consider the euclidean space of vectors $L\left(\frac{L_1}{L_1^0}, \frac{L_2}{L_2^0}, \ldots, \frac{L_k}{L_k^0}\right)$. The sum (6.11) represents the square of the distance from the arbitrary point $X \in \Omega$, corresponding to a plan, to the utopian point having coordinates (1,1,...1). The problem becomes that of selecting a vector optimal plan $X^* \in \Omega$, which minimizes this distance.

A Physical Interpretation of the Problem. If we are given a vector optimal plan $X^*(x_1^*,\ldots,x_n^*)$, then this plan will be that for which the numbers $L_1(X^*), L_2(X^*),\ldots,L_k(X^*)$ will be as near as possible to $L_1^0, L_2^0,\ldots,L_k^0$, respectively.

Here is the meaning of this approximation: we assume that we have chosen any of the linear forms $L_\alpha(X)$ from (6.8) and have optimized it over the region Ω. Then we take number $L_\alpha(X_\alpha^0)$ as the performance measure of the system, which we may achieve under optimization of each individual form from (6.8), taken separately. Now we try to improve several performance indices simultaneously. Consequently, under the choice $X^* \in \Omega$, there is a deterioration of several individual performance indices; however, this deterioration is distributed over the entire set of indices (6.8) and is the minimal possible deterioration.

We note that the physical meaning of the given problem may be expressed by a model, analogous to the presentation in (174).

Remark 1. The sum (6.11) is a dimensionless function. It is the square of the distance in a euclidean space of the difference between the dimensionless vector

$$L\left(\frac{L_1}{L_1^0}, \frac{L_2}{L_2^0}, \ldots, \frac{L_k}{L_k^0}\right)$$

and the unit vector.

Remark 2. In the case when any of the values (6.10) turn out to be zero, we construct a transformation of the region Ω so that once again the values (6.10) will all be different from zero.

Remark 3. The problem may also be interpreted as an approximate solution to the inconsistent set of linear algebraic equations (145)

$$\sum_{j=1}^{n} \frac{c_{ij}}{L_i^0} x_j - 1 = 0 \qquad (i=1,2,\ldots,k).$$

Remark 4. Clearly, k linear forms may be treated in the usual way, solving the linear programming problem for the sum $\sum_{\alpha=1}^{k} \lambda_\alpha L_\alpha(X)$, in which $\lambda_1,\lambda_2,\ldots,\lambda_k$ are numerical multipliers, reducing the situation, via the given sum, to a single dimensional problem. Such an approach to the problem is possible only in those cases when the numerical values of the coefficients are known exactly in advance.

For nonlinear programming problems, the ideas of the above-stated formulation of the vector optimization problem retain the same meaning. However, in such cases instead of the linear forms (6.8), we will have the vector function G(X) with components $g_1(X),\ldots,g_k(X)$, while the linear constraints (6.2) - (6.4) are replaced by the conditions (6.5). Consequently, instead of (6.11) we will have

$$R(X) = \sum_{\alpha=1}^{k} \left(\frac{g_\alpha(X)}{g_\alpha^0} - 1 \right)^2, \tag{6.13}$$

where g_α^0 is the minimal (maximal) value of the function $g_\alpha(X)$ in Ω under (6.5).

6. Determination of the Vector-Optimal Plan

The determination of the vector-optimal plan $X^* \in \Omega$ in the linear programming problem has been reduced to the minimization of the expression (6.11) over solutions of the system of linear inequalities (6.2) - (6.4).

Since (6.11) is a quadratic expression in the variables x_1, $x_2,\ldots,x_n$, the problem of determination of the vector-optimal plan is reduced to a convex programming problem or, more precisely, to a quadratic programming problem.

Given a convex function R(X), defined on the set $X \in \Omega$, we are required to determine a point $X^* \in \Omega$ satisfying the condition

$$R(X^*) = \min_{X\in\Omega} R(X).$$

As is known, in convex programming problems a local minimum coincides with the absolute minimum, i.e. with the smallest value of the function in the domain Ω. Therefore, for the solution of the convex programming problem we may use any methods which leads to a local extremum, including gradient methods (33,175-177), as well as the methods based upon the electronic models discussed in (178,179). Several algorithms for the solution of convex and quadratic programming problems are described in detail in (173).

The determination of the vector-optimal plan $X^* \in \Omega$ in non-linear programming problems has been reduced to the minimization of (6.13) under the conditions (6.5). For determination of X^*, we may apply search methods allowing us to find extremal points of complex multidimensional functions.

7. Compromise Solutions

In vector optimization problems, for choice of a solution a natural first step is to consider identifying the domain of compromise solutions, i.e. Pareto-optimal or, as it was called above, the unimprovable solutions. In Chapter I, we mentioned that for dynamical systems several results in this direction were obtained in the works (54-58) and for static problems in (64-71,90-99, 106-128).

The regions of compromise solutions $\Gamma \in \Omega$ is usually a subset Γ of the admissible convex set $\Omega \subset E^n$, possessing the property that each solution lying in Γ cannot simultaneously improve all the scalar criteria. This means that the values $G(X')$ and $G(X'')$ of the vector goal $G(X)$ are in conflict for at least one component for any two solutions $X' \varepsilon \Gamma$, $X'' \varepsilon \Gamma$.

Naturally, any optimal solution in a vector optimization problem must lie in the region of compromise solutions Γ. If not, then the solution may be improved and, consequently, it is not optimal. As is well known (cf. (134)) and was shown above, in many cases the region of compromise solutions Γ may be obtained by optimizing a linear combination of components of the vector $G(X)$, i.e.

$$\sum_{\alpha=1}^{k} \lambda_\alpha g_\alpha(X), \tag{6.14}$$

with the non-negative coefficients $\lambda_1, \lambda_2, \ldots, \lambda_k$ taking on any values such that

$$\sum_{\alpha=1}^{k} \lambda_\alpha = 1. \tag{6.15}$$

Definition of the domain of compromise solutions essentially narrows down the search for optimal solutions; however, the difficulty in determination of such solutions remains since to establish the corresponding best solution vectors $\lambda\{\lambda_1, \lambda_2, \ldots, \lambda_k\}$ is almost impossible in all practical cases. The task of determining the vector λ is done by a principle of compromise in most problems.

In (91-99) it is proposed to separate the dominant solutions from the set of Pareto-optimal solutions, i.e. the compromise solutions. The principle of separation of such solutions is to minimize a measure of approximation, such as in (2.11), in the space of optimal criteria. For studying this approach, we write it in the form

$$R_1(G(X)) = \left[\sum_{\alpha=1}^{k} (g_\alpha(X) - g_\alpha^0)^\ell\right]^{\frac{1}{\ell}}, \quad \ell \geq 1. \tag{6.16}$$

As already noted in section 2 of Chapter II, minimization of (6.16) for $\ell = 1$ is equivalent to minimization of (6.14) with coefficients $\lambda_1=\lambda_2=\ldots=\lambda_k > 0$; for $\ell = 2$, (6.16) is the distance of the variables $g_1, g_2, \ldots, g_k$ from an arbitrary point to a "utopian" point with coordinates $g_1^0, g_2^0, \ldots, g_k^0$, while for $\ell = \infty$ it represents the estimate

$$R_\infty(G(X)) = \max_\alpha \{(g_\alpha(X) - g_\alpha^0) \mid \alpha=1,\ldots,k\},$$

the minimization of which gives the maximal value of all criteria deviations from their optimal guaranteed smallest values.

The dominant solutions are determined as the set $\Gamma_\ell \subset \Gamma \subset \Omega$, of points X_ℓ, determined by minimizing (6.16) for $\ell \in [1,\infty]$. Each solution $X_\ell \in \Gamma_\ell$ is called a compromise solution with parameter ℓ.

Usually the space Ω is called the space of solutions, while the space of variables $g_1(X),\ldots g_k(X)$ is the space of criteria.

In the space of criteria we consider the set

$$G_\Omega = \{G(X) \mid X \in \Omega\}.$$

We note that if G_Ω is a compact set, then dominant solutions X_ℓ exist for any $\ell \geq 1$ and we always have Pareto-optimal solutions for $1 \leq \ell < \infty$. In particular, in (101) it is proved that X_2 is a Pareto-optimal solution.

We show that the dominant solution X_ℓ is actually the compromise solution (Pareto-optimal) for any $\ell \in [1,\infty)$, i.e. $X_\ell \in \Gamma$.

We have

1) $X \in \Omega$, Ω convex in E^n;
2) $\forall X \in \Omega$: $g_\alpha^0 = g_\alpha(X^0) \leq g_\alpha(X)$ $(\alpha=1,\ldots,k)$;

3) the set of dominant solutions

$$\Gamma_\ell = \left\{ X_\ell \,|\, X_\ell \in \Omega; \quad \forall X \in \Omega, \quad \forall \ell \in [1,\infty): \right.$$

$$\left. \left[\sum_{\alpha=1}^{k} (g_\alpha(X_\ell) - g_\alpha^0)^\ell\right]^{1/\ell} \le \left[\sum_{\alpha=1}^{k} (g_\alpha(X) - g_\alpha^0)^\ell\right]^{1/\ell} \right\};$$

4) the set of compromise solutions

$$\Gamma = \{\overline{X} \,|\, \overline{X} \in \Omega; \quad \forall X \in \Omega:$$

$$g_\alpha(\overline{X}) \le g_\alpha(X) \;\; (\alpha=1,\ldots,m<k), \quad g_\alpha(\overline{X}) < g_\alpha(X) \;\; (\alpha=m+1,\ldots,k)\}.$$

We assume the contrary: X_ℓ does not lie in the set Γ. Then we find an $X' \in \Omega$ such that

$$g_\alpha(X') \le g_\alpha(X_\ell) \qquad (\alpha=1,\ldots,m<k),$$

$$g_\alpha(X') < g_\alpha(X_\ell) \qquad (\alpha=m+1,\ldots,k), \qquad (6.17)$$

$$g_\alpha(X') \ge g_\alpha^0 \qquad (\alpha=1,\ldots,k).$$

The last inequality of (6.17), using the first two inequalities, may be written inthe form

$$0 \le g_\alpha(X') - g_\alpha^0 \le g_\alpha(X_\ell) - g_\alpha^0 \qquad (\alpha=1,\ldots,m<k),$$

$$0 \le g_\alpha(X') - g_\alpha^0 < g_\alpha(X_\ell) - g_\alpha^0 \qquad (\alpha=m+1,\ldots,k). \qquad (6.18)$$

From (6.18), we have

$$\left[\sum_{\alpha=1}^{k} (g_\alpha(X') - g_\alpha^0)^\ell\right]^{1/\ell} < \left[\sum_{\alpha=1}^{k} (g_\alpha(X_\ell) - g_\alpha^0)^\ell\right]^{1/\ell}, \tag{6.19}$$

contradicting the definition of the dominant solution X_ℓ (point 3). Consequently, $X_\ell \in \Gamma$.

The dominant solution X_ℓ, with $\ell = \infty$, may not be Pareto-optimal since in this case it is not guaranteed that (6.19) will be satisfied.

Choice of the parameter ℓ and, by the same token, identification of concrete solutions from Γ, requires additional discussion.

In (134) a strategy of true compromise is given based upon introduction of a measure of relative reduction of the quality of the solution with respect to each of the criteria. For example, in the case of a vector index $G\{g_1,g_2\}$, from the two compromise solutions $X', X'' \in \Gamma$, the preference is given to that solution corresponding to the "value concession" which is greater. This means that X' is preferred to X" if

$$\frac{|g_1(X') - g_1(X'')|}{\max\limits_{X',X''} g_1(X)} > \frac{|g_2(X') - g_2(X'')|}{\max\limits_{X',X''} g_2(X)},$$

and conversly.

In (66) the special case of a continuous game is studied and it is shown that the game strategy, based on optimization of (6.16) for $\ell = 2$, is preferable to those strategies such as "Nash equilibrium," "minimax" and "Pareto-optimal," generally speaking.

In our view, compromise solution, based on minimization of (6.11), is logical only in those cases when there is an absence of complete information about choice of the coefficients λ_α in (6.15).

8. The Existence of a Vector-Optimal Plan

We do not give here the well known Kuhn-Tucker theorem (24,51) on the existence of a solution to nonlinear programming problems, but establish only the existence of a vector-optimal plan for linear programming problems with vector criteria of the type considered above in section 5.

We have the linear constraints (6.2) - (6.4), defining the space of solutions $\Omega \ni X$ and the system of linear forms (6.8). The space of solutions Ω in this case is a convex polyhedron in the nonnegative orthant. The values L_i^0 will be assumed at the vertices of this convex polyhedron. For a definite measure of approximation $R(X)$ as a smooth, convex function of the vector $X \varepsilon \Omega$, the solution of the problem will exist and is situated either on the boundary or interior of the polyhedron Ω. This assertion remains valid for cases when Ω is a convex subspace of the space of non-negative variables $x_1, x_2, \ldots, x_n$.

Since for $\ell = 2$, (6.16) is a convex function, the compromise solution will be situated either on the boundary or in the interior of Ω. For $\ell = 1$, (6.16) turns into a linear form and the compromise solution in this case will be assumed on the boundary of the convex polyhedron Ω.

9. A Numerical Example of the Determination of a Vector-Optimal Plan

In this section we consider a numerical example of the determination of a vector-optimal plan in the linear programming problem with multiobjectives. In the following section we consider a concrete planning problem with definite physical meaning.

Let there be given the system of linear forms

$$L_1(X) = -3x_1 + 2x_2,$$
$$L_2(X) = 4x_1 + 3x_2, \qquad (6.20)$$
$$L_3(X) = 2x_1 - 5x_2$$

and the inequalities

$$-2x_1 - 3x_2 + 18 \geq 0,$$
$$-2x_1 - x_2 + 10 \geq 0, \qquad (6.21)$$
$$x_1 \geq 0, \quad x_2 \geq 0,$$

defined in a region Ω of admissible values of x_1, x_2.

We determine a plan $X^*(x_1^*, x_2^*) \in \Omega$, under which the forms (6.20) assume their maximal values, simultaneously.

Applying the simplex method (173), we determine the maximal values which these forms, may assume on the solutions of the inequalities (6.21).

The linear form $L_1(X)$ assumes its maximal value at the point $x_1 = 0$, $x_2 = 6$ and equals $L_1^0 = 12$. In this case, $L_2 = 18$, $L_3 = -30$.

The linear form $L_2(X)$ takes on its maximal value at $x_1 = 3$, $x_2 = 4$ and equals $L_2^0 = 24$. At this point, $L_1 = -1$, $L_3 = -14$.

The linear form $L_3(X)$ is maximized at $x_1 = 5$, $x_2 = 0$ and equals $L_3^0 = 10$. At this point, $L_1 = -15$, $L_2 = 20$.

We form the function $R(X)$. According to (6.11), we have

$$R(x_1, x_2) = \left[12 - (-3x_1 + 2x_2)\right]^2 + \left[24 - (4x_1 + 3x_2)\right]^2 + \left[10 - (2x_1 - 5x_2)\right]^2,$$

which, after some algebra, is

$$R(x_1,x_2) = 29x_1^2 + 38x_2^2 - 8x_1x_2 - 160x_1 - 92x_2 + 820.$$

The minimal value of this function is assumed at the point $X^*(x_1^*,x_2^*)$ with coordinates

$$x_1^* = \frac{1612}{543} \approx 2.97; \qquad x_2^* = \frac{827}{543} \approx 1.52, \tag{6.22}$$

which lies in the set Ω of solutions to the inequalities (6.21) and is the vector-optimal plan for the problem under study. For this plan, the optimized linear forms take on the values

$$L_1^* = -5.87; \quad L_2^* = 16.44; \quad L_3^* = -1.66$$

For comparison, we compute several representative plans of interest:

a) for maximization of the sum of the linear forms (6.20),

$$L = L_1 + L_2 + L_3 = 3x_1$$

the plan will be $x_1 = 5$, $x_2 = 0$, coinciding with the plan obtained by maximizing only the form $L_3(X)$.

b) We compute the Chebyshev point for the system of linear equations

$$\eta_1(x_1,x_2) = -3x_1 + 2x_2 - 12 = 0,$$

$$\eta_2(x_1,x_2) = 4x_1 + 3x_2 - 24 = 0, \tag{6.23}$$

$$\eta_3(x_1,x_2) = 2x_1 - 5x_2 - 10 = 0$$

under the constraints (6.21). It is determined by minimizing the function

TABLE 5

Conditions of the Problem: Linear Forms and Constraints	Problem Plan	Result of the Action	Value of L_1	Value of L_2	Value of L_3
	$x_1 = 0$, $x_2 = 6$	$\max(-3x_1 + 2x_2)$	12	18	-30
	$x_1 = 3$, $x_2 = 4$	$\max(4x_1 + 3x_2)$	-1	24	-14
$L_1 = -3x_1 + 2x_2$ $L_2 = 4x_1 + 3x_2$ $L_3 = 2x_1 - 5x_2$	$x_1 = 5$, $x_2 = 0$	$\max(2x_1 - 5x_2)$	-15	20	10
	$x_1 = 5$, $x_2 = 0$	$\max(L_1 + L_2 + L_3) = \max x_1$	-15	20	10
$-2x_1 - 3x_2 + 18 \geq 0$ $-2x_1 - x_2 + 10 \geq 0$	$x_1 = 1.52$, $x_2 = 1.37$	$\min V(X)$ $V(X) = \max \lvert L_1 - 12, L_2 - 24, L_3 - 10 \rvert$	-1.82	10.19	-3.81
$x_1 \geq 0$ $x_2 \geq 0$	$x_1 = 5$, $x_2 = 0$	$\min U(X)$ $U(X) = \lvert L_1 - 12\rvert + \lvert L_2 - 24\rvert + \lvert L_3 - 10\rvert$	-15	20	10
	$x_1^* = 2.97$, $x_2^* = 1.52$	$\min R(X)$ $R(X) = (L_1 - 12)^2 + (L_2 - 24)^2 + (L_3 - 10)^2$	-5.87	16.44	-1.66

$$V(x_1,x_2) = \max_{1 \leq i \leq 3} |\eta_i(x_1,x_2)|.$$

Applying the simplex method (173), we find the point

$$x_1 = \frac{82}{54} \approx 1.52; \quad x_2 = \frac{518}{378} \approx 1.37,$$

which gives the linear forms (6.20) the following values:

$$L_1 = -1.82; \quad L_2 = 10.19; \quad L_3 = -3.81.$$

c) We calculate the plan which minimizes the modulus of the linear functions (6.23)

$$U(x_1,x_2) = \sum_{i=1}^{3} |\eta_i(x_1,x_2)|.$$

Using the simplex method (173), we find the minimizing point

$$x_1 = \frac{2550}{442} \approx 5.77; \quad x_2 = \frac{136}{442} \approx 0.31.$$

The point obtained is situated at the intersection of the lines $L_1(X) = 24$, $L_2(X) = 10$ and doesn't satisfy the problem inequalities (6.21). Under the constraints (6.21), the simplex method determines the plan of the problem as $x_1 = 5$, $x_2 = 0$, again coinciding with the plan obtained by maximizing only $L_3(X)$.

For completeness, all the results are summarized above in Table 5. It is not difficult to observe that, of all the plans, only the vector-optimal plan (6.22) provides the closest approximation of the linear forms (6.20) to their maximal attainable values in the region (6.21).

10. A Planning Problem in the Metallurgical Industry

We consider a simplified version of a concrete planning problem for the metallurgical industry.

We assume that a ferro alloy plant over a definite time interval has the goal of smelting n tons of a silicon-manganese alloy and has the possibility of ordering ore (a mixture of manganese and carbon-manganese concentrate) from mines A and B situated at different locations from the smelter at distances ℓ_A and ℓ_B, km, respectively. The quantity of ore ordered by the smelter from these mines we denote as x_A and x_B, respectively.

Let the ore from mine A have the following characteristics: for 1 ton of ore, the smelter must pay c_A rubles; for its processing t_A kw-hrs/ton of electric energy is required; it contains m_A% phosphorus; from it is produced n_A tons of silicon-manganese alloy. We have the analogous parameters for ore from mine B as: c_B[rubles/ton], t_B[kw-hrs/ton], m_B[%], n_B[tons], respectively.

We assume that the smelter capacity over the given time interval is such that n_1 tons of ore can definitely be processed. Moreover, the smelter is constrained in electric energy and it is unprofitable to spend more than t kw-hrs/ton of electrical energy for smelting the alloy. The quantity of electrical energy expended to smelt one ton of alloy may be written as

$$\frac{t_A x_A + t_B x_B}{n_A x_A + n_B x_B} .$$

According to the given technology, the per cent of phosphorus in the alloy is

$$\frac{m_A x_A + m_B x_B}{n_A x_A + n_B x_B}$$

which must not exceed a given amount m.

For such a problem, the planning must be carried out with regard to constraints expressed by the inequalities

$$
\begin{aligned}
&n_A x_A + n_B x_B \geq n, \\
&x_A + x_B \leq n_1, \\
&t_A x_A + t_B x_B \leq t(n_A x_A + n_B x_B), \\
&m_A x_A + m_B x_B \leq m(n_A x_A + n_B x_B), \\
&x_A \geq 0, \quad x_B \geq 0.
\end{aligned}
\tag{6.24}
$$

These inequalities may be used as conditions for a linear programming problem. The first inequality insures smelting the planned quantity of alloy, the second expresses the capability of the smelter for processing the ore, the third restricts the amount of electrical energy to be within the admissible limit, while the fourth insures that the required limit on phosphorus per cent in the alloy will be met.

In the (x_A, x_B)-plane, the inequalities (6.24) define the region Ω.

The value of the smelted ore is determined by the linear form $c_A x_A + c_B x_B$, the relation of which to the amount of smelted alloy $n_A x_A + n_B x_B$ is determined by the average cost of the raw material needed for smelting one ton of alloy. For sale of the smelted alloy, the smelter obtains a definite return.

The sense of the plan is, on the one hand, to purchase the ore so that the overall return given by

$$L_1(x_A, x_B) = c(n_A x_A + n_B x_B) - c_A x_A - c_B x_B - p(\ell_A x_A + \ell_B x_B), \tag{6.25}$$

is maximal, where c denotes the value of one ton of alloy, while p expresses the cost of transporting one ton of ore a distance of 1 km.

On the other hand, planning the purchase of the ore may be complicated by a requirement of a minimal run of transportation means for supplying the ordered ore. This requirement may be given by the total number of ton-kilometers required by the smelter for transportation of the ore, and is expressed in the form

$$L_2(x_A, x_B) = \ell_A x_A + \ell_B x_B. \tag{6.26}$$

We obtain a linear programming problem with two objective functions--the planning performance indices (6.25) and (6.26). Both indices should be optimized.

The linear programming problem consists in the simultaneous optimization of the planning performance indices (6.25), (6.26) under the constraints (6.24).

We solve the planning problem under concrete values of the parameters, close to those used in actual operation. Let n = 350 tons, n_1 = 1000 tons, n_A = 0.5 tons, n_B = 0.3 tons, t = 4000kw-hrs/ton, t_A = 1500 kw-hrs/ton, t_B = 1300 kw-hrs/ton, c_A = 32.5 rubles/ton, c_B = 12.5 rubles/ton, c = 150 rubles/ton, m_A = 0.18%, m_B = 0.17%, m = 0.5%, p = 0.1 rubles/ton-km, ℓ_A = 50 km, ℓ_B = 25km. (The value of c is given without regard for the remaining smelter expenses. Naturally, it will be larger. Here the general planning problem has been simplified.)

Consequently, the linear programming problem takes on the concrete form

$$\begin{aligned} &0.5x_A + 0.3x_B \geq 350, \\ &x_A + x_B \leq 1000, \\ &x_A \geq 0, \quad x_B \geq 0 \end{aligned} \tag{6.27}$$

where we seek the maximal and minimal values of the linear forms

$$L_1(x_A,x_B) = 37.5x_A + 30x_B, \tag{6.28}$$

$$L_2(x_A,x_B) = 50x_A + 25x_B \tag{6.29}$$

respectively.

The inequalities (6.27) determine the region Ω in the (x_A,x_B)-plane and represent the triangle abc in Fig. 11.

It is not difficult to compute that the linear form (6.28), by itself, takes on its maximal value for

$$x_A = 1000T, \quad x_B = 0,$$

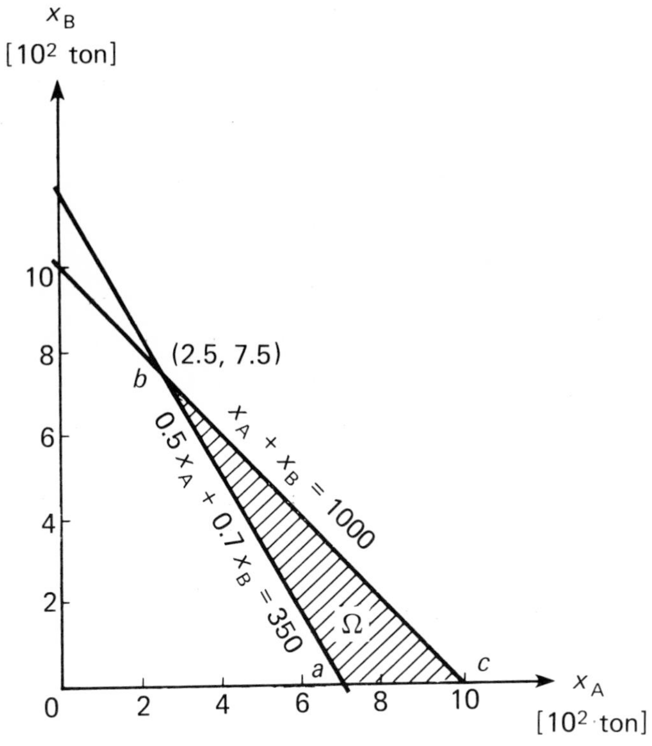

Fig.11

for which the linear forms (6.28) and (6.29) assume the values

$$L_1^0 = 37{,}500 \text{ rubles}, \quad L_2 = 50{,}000\text{T/km}.$$

respectively.

Under this plan, for smelting one ton of alloy the purchase and transport of ore requires $\frac{c_A + p\ell_A}{n_A} = 75$ rubles, and 500 tons of alloy is smelted.

The linear form (6.29), by itself, is minimized at the point b (Fig. 11); its coordinates are

$$x_A = 250\text{T}; \quad x_B = 750\text{T}$$

where the linear forms (6.28), (6.29) assume the values

$$L_1 = 31{,}875 \text{ rubles}, \quad L_2^0 = 31{,}250\text{T/km}.$$

In this case, for one ton of alloy, the purchase and transport costs are 58 rubles and 98 kopecks but the overall gain is reduced at the expense of a reduction in the quantity of smelted alloy of 350 tons.

Thus, an increase in profit of 5625 rubles is accompanied by an increase of transportation resources by an amount 18750 ton-kilometer.

We determine the vector-optimal plan for the problem.

We form the function

$$R(X) = \left(\frac{37.5x_A + 30x_B}{37{,}500} - 1\right)^2 + \left(\frac{50x_A + 25x_B}{31{,}250} - 1\right)^2$$

and in the domain Ω, defined by the inequalities (6.23), we find the value of the variables x_A^*, x_B^*, minimizing this function.

Computations in the region Ω determine the point

$$x_A^* = 294T; \quad x_B^* = 706T, \tag{6.30}$$

situated on the segment bc (Fig. 11).

Under such a plan the problem objective functions take on the values

$$L_1^* = 32{,}205 \text{ rubles}; \quad L_2^* = 32{,}350T/\text{km}.$$

For completeness, the results of the calculations are presented in Table 6.

The obtained vector-optimal plan (6.30) optimizes the profit increase in its connection with the growth of transportation resources required. Under the considered level of ore processing, the optimal is considered to be an increase in transportation resources of 1100 ton-kilometer, which is accompanied by an increase of overall profit of 330 rubles. In such a case, for smelting one ton of alloy, the purchase and transport of ore costs 60 rubles, 24 kopecks, and the final product smelted amounts to 358.8 tons.

Under the plan (6.30), for the per cent phosphorus in the alloy and the amount of electrical energy, attention should be paid not just to keeping them within the admissible limits. Naturally, a decrease in the per cent of phosphorus in the alloy, i.e. an increase in the quality of the alloy, and also a decrease in the amount of electrical energy, would raise the overall return.

Accounting for these requirements complicates the planning problem but it remains within the scope of the method presented. Also, in principle the problem may be solved taking into consideration other requirements of both a technological, as well as organizational character.

TABLE 6

Problem Conditions	Problem Plan	Function to be Minimized
$x_A + x_B \leq 1000$ $0.5x_A + 0.3x_B \geq 350$ $x_A \geq 0$ $x_B \geq 0$	$x_A = 1000T$ $x_B = 0$	$L_1 = -(37.5x_A + 30x_B)$
	$x_A = 250T$ $x_B = 750T$	$L_2 = 50x_A + 25x_B$
	$x_A^* = 294T$ $x_B^* = 706T$	$R = \left(\frac{37.5x_A + 30x_B}{37{,}500} - 1\right)^2 + \left(\frac{50x_A + 25x_B}{31{,}250} - 1\right)^2$

It is possible to approach the planning process from different points of view, depending on different factors and requirements. Here we consider a concrete example.

We approach the planning problem from the viewpoint of the minimization of the cost of the ore for smelting 1 ton of alloy, so that the per cent phosphorus in the alloy not only raises the chances for realization of the product, but also increases the overall return. We assume that the smelting of 1 ton of ore, together with its transport from the mines A and B costs c_A and c_B rubles, respectively. Also, we assume that the mines A and B guarantee supply of only $\bar{x}_A$ and $\bar{x}_B$ tons over the planning period. The remaining parameters are as in the example above.

Total Profit L_1 rubles	Transport Costs L_2 T/km	Expense for Purchase and Transport of Ore for Smelting 1 ton rubles	Quantity of Alloy T
37,500	50,000	75	500
31,875	31,250	58.93	350
32,205	32,350	60.24	358.8

Such a planning problem must be solved with due regard for the constraints expressed as

$$\begin{aligned} &n_A x_A + n_B x_B \geq n, \\ &t_A x_A + t_B x_B \leq t(n_A x_A + n_B x_B), \\ &m_A x_A + m_B x_B \leq m(n_A x_A + n_B x_B), \\ &0 \leq x_A \leq \bar{x}_A, \quad 0 \leq x_B \leq \bar{x}_B. \end{aligned} \tag{6.31}$$

These inequalities may be used as conditions for a linear programming problem. As in the preceding problem, the first inequality ensures smelting of the planned amount of alloy, the second contains the amount of electrical energy, while the third guarantees satisfaction of the requirement on phosphorus content in the alloy. The inequalities (6.31) determine a region Ω in the (x_A, x_B)-plane.

We consider the case when the smelter plans to smelt n tons of alloy and it is expedient to plan the purchase of ore so that the expense is minimal, while to smelt the highest quality silicon-manganese possible from the purchased ore. The latter condition is determined by the percentage of phosphorus in the alloy; the less phosphorus in the alloy, the better the alloy. Such an approach to planning may be carried out not only from the viewpoint of maximal return from the sale of quality alloy, but also from the essential risk involved in its realization. The smelter may smelt a lower quality alloy for minimal expense, while it may not be able to realize it. On the other hand, the smelter guarantees its realization for a high-quality alloy, but the expense for the acquisition of raw materials may be unacceptable. These requirements may be expressed by the linear forms

$$L_1(x_A, x_B) = c_A x_A + c_B x_B, \tag{6.32}$$

$$L_2(x_A, x_B) = 0.01 m_A x_A + 0.01 m_B x_B. \tag{6.33}$$

The form (6.32) determines the overall cost of the ore purchased (in rubles), while the form (6.33) is the amount of phosphorus (in tons) in the produced alloy. Here we make the assumption that under smelting, the weight of phosphorus in the ore is transferred to the alloy.

We obtain a linear programming problem with two objective functions: to find a solution of (6.31) which minimizes (6.32) and (6.33) simultaneously.

We consider the problem for actual values of the parameters, which are near to those used in practice. Let n = 15,000 tons, n_A = 0.6 tons, n_B = 0.3 tons, t = 4000 kw-hrs/ton, t_A = 1650 kw-hrs/ton, t_B = 1450 kw-hrs/ton, c_A = 52.5 rubles/ton, c_B = 15 rubles/ton, m_A = 0.174%, m_B = 0.19%, $\bar{x}_A$ = 300,000 tons, $\bar{x}_B$ = 400,000 tons, m = 0.5%.

For these values, (6.31) assumes the form

$$
\begin{aligned}
&2x_A + x_B \geq 500{,}000,\\
&3x_A - x_B \geq 0,\\
&3.14x_A - x_B \geq 0,\\
&0 \leq x_A \leq 300{,}000, \quad 0 \leq x_B \leq 400{,}000,
\end{aligned}
\tag{6.34}
$$

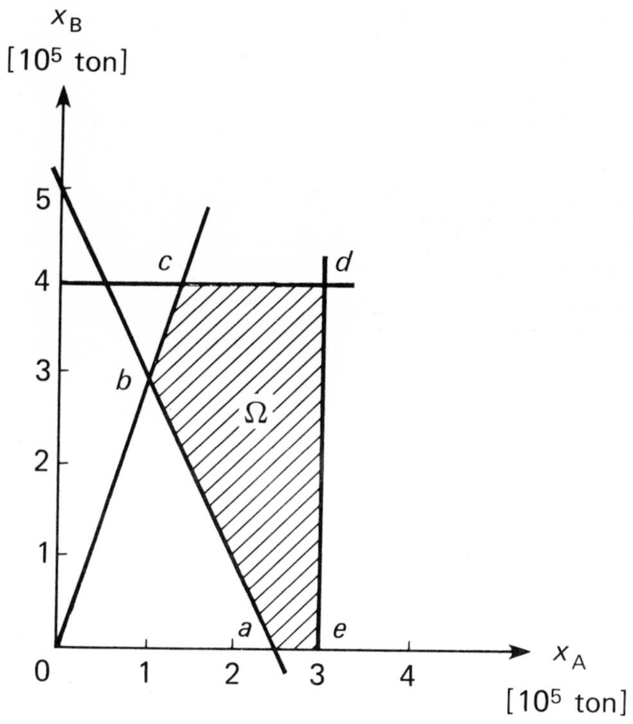

Fig. 12

while the linear forms (6.32) and (6.33) can be written as

$$L_1(x_A, x_B) = 52.5x_A + 15x_B, \tag{6.35}$$

$$L_2(x_A, x_B) = 0.00174x_A + 0.0019x_B, \tag{6.36}$$

respectively.

The region Ω determined by the inequalities (6.34) is shown in Fig. 12.

For the case considered of smelting only the planned amount $n = 150{,}000$ tons of alloy, the constraints (6.34) are written as

$$\begin{gathered} 2x_A + x_B = 500{,}000, \\ 3x_A - x_B \geq 0, \\ x_A \geq 0, \quad x_B \geq 0, \end{gathered} \tag{6.37}$$

which shows that the solution of the posed problem should be sought on the segment ab (Fig. 12).

The optimal plan of the problem of minimizing only the cost of the ore purchased, is easily computed and is

$$x_A = 100{,}000\text{T}; \quad x_B = 300{,}000\text{T}.$$

Under such a plan, the linear forms (6.35) and (6.36) take the values

$$L_1^0 = 9{,}750{,}000 \text{ rubles}; \quad L_2 = 744\text{T},$$

with the help of which we determine the average cost of the ore for smelting 1 ton of alloy as 65 rubles and the per cent of phosphorus in the alloy is 0.496%, which is within the admissible limits.

The optimal plan of the problem of minimizing only the phosphorus content in the alloy is

$$x_A = 250{,}000T; \quad x_B = 0.$$

Under this plan, the linear forms (6.34) and (6.35) assume the values

$$L_1 = 13{,}125{,}000 \text{ rubles}, \quad L_2^0 = 435T,$$

and the average cost of the ore for smelting 1 ton of alloy and the phosphorus percentage in the alloy will be 87.5 rubles and 0.29%, respectively.

Under the given problem conditions, it is impossible to smelt alloy with a smaller per cent of phosphorus.

As is seen from these figures, a decrease in phosphorus content from 0.496% to 0.29% is connected with an increase in expenditure for ore purchase of more than 3 million rubles.

We determine the vector optimal plan of the problem. We form the function

$$R(X) = \left(\frac{52.5x_A + 15x_B}{9{,}750{,}00} - 1\right)^2 + \left(\frac{0.00174x_A + 0.0019x_B}{435} - 1\right)^2$$

and in the domain defined by (6.37), we seek values x_A^*, x_B^* minimizing this function. They are computed to be

$$x_A^* = 221{,}217T; \quad x_B^* = 57{,}566T. \tag{6.38}$$

The results of the entire computation are presented in Table 7.

TABLE 7

Problem Conditions	Problem Plan	Functional to be Minimized
	$x_A = 100{,}000T$ $x_B = 300{,}000T$	$L_1 = 52.5x_A + 15x_B$
$2x_A + x_B = 500{,}000$ $3x_A - x_B \geq 0$ $x_A \geq 0$ $x_B \geq 0$	$x_A = 250{,}000T$ $x_B = 0$	$L_2 = 0.00174x_A + 0.0019x_B$
	$x_A^* = 221{,}217T$ $x_B^* = 57{,}566T$	$R = \left(\frac{52.5x_A + 15x_B}{9{,}750{,}000} - 1\right)^2 + \left(\frac{0.00174x_A + 0.0019x_B}{435} - 1\right)^2$

The vector-optimal plan (6.38) of the posed problem to some degree minimizes both the cost of the ore purchased, which is needed for smelting 1 ton of alloy, and the per cent phosphorus content in the alloy, which, in turn maximally decreases the risk of realization of the smelted product.

Under the vector-optimal plan (6.38), the objective functions (6.35), (6.36) take on the values

$$L_1^* = 12{,}477{,}382.5 \text{ rubles}; \quad L_2^* = 494.293T.$$

Value L_1 rubles	Value L_2 T	Cost of the Ore for Smelting 1 Ton of Alloy rubles	Per Cent Phosphorus in the Alloy %
9,750,000	744	65	0.496
13,125,000	435	87.5	0.29
12,477,382.5	494,293	83.18	0.33

For such a plan, the mean cost of the ore needed for smelting 1 ton of alloy will be 83.18 rubles, while the per cent phosphorus in the alloy will be 0.33%.

The meaning of the planning process, based on the vector-optimal plan, is that the optimal operation of the smelter turns out to be that which increases the phosphorus per cent from the minimization possible by 0.4%. At the same time, this action decreases the expense of smelting each ton of alloy by 4.32 rubles.

CHAPTER VII

PARAMETER OPTIMIZATION IN ENGINEERING SYSTEMS

1. Problem Statement

The general formulation of the mathematical problem of determining optimal parameters in engineering systems is almost the same as the formulation of the nonlinear programming problem given in section 2 of Chapter VI. However, here there is a specific character to the optimizing function $g(X)$ and to the set Ω of inequalities (6.5). These functions have the form of posynomials or combinations of posynomials. A posynomial is a function of n variables $x_1, x_2, \dots, x_n$ having the form (26):

$$g(X) = \sum_{i=1}^{m} b_i x_1^{\alpha_{i1}} x_2^{\alpha_{i2}} \dots x_n^{\alpha_{in}}, \tag{7.1}$$

where $x_i > 0$, $b_i > 0$, and α_{ij} are arbitrary real numbers ($j=1,\dots,n$; $i=1,\dots,m$).

A simplified problem of determining optimal parameters of engineering systems is mathematically formulated in the following form: to find a vector $X^0(x_1^0, x_2^0, \dots, x_n^0)$ under which the function $g_0(X)$ is minimized and which satisfies the constraints

$$x_1 \geq 0, \quad x_2 \geq 0, \dots, x_n \geq 0, \tag{7.2}$$

$$g_1(X) \leq 0, \quad g_2(X) \leq 0, \dots, g_p(X) \leq 0, \tag{7.3}$$

where the functions $g_0(X), g_1(X), \dots, g_p(X)$ have the posynomial forms

$$g_k(X) = \sum_{i=1}^{m_k} b_i^{(k)} x_1^{\alpha_{i1}^{(k)}} x_2^{\alpha_{i2}^{(k)}} \ldots x_n^{\alpha_{in}^{(k)}} \tag{7.4}$$

$(k=0,1,\ldots,p)$.

Here $m_0, m_1, \ldots, m_k$ are positive integers, while $b_i^{(k)} > 0$; $\alpha_{ij}^{(k)}$ are arbitrary real numbers.

A more complicated version of the above problem is when the functions $g_0(X), g_1(X), \ldots, g_p(X)$ are combinations of posynomials like (7.1).

The solution of the presented problem may naturally be effected by general nonlinear programming methods (19-24,26). In (26), nonlinear programming methods are developed for application to problems of the above type, considering a specific function and posynomial constraints of the type (7.1). These methods are termed geometric programming methods.

A determination of the parameters of an engineering system minimizing only the function $g_0(X)$ is unacceptable in many cases since attention is not paid to other, no less important, performance indices of the system. For an effective solution of the problem, we present the engineering system problem as a vector optimization problem: we wish to determine a vector $X^*(x_1^*, x_2^*, \ldots, x_n^*)$ for which the functions $f_1(X), f_2(X), \ldots, f_k(X)$ take on their minimal values and for which the constraints (7.2) and (7.3) are satisfied. We assume the functions $f_\alpha(X)$, $g_\beta(X)$, $(\alpha=1,\ldots,k;\ \beta=1,\ldots,p)$ are represented as posynomials of the type (7.1) or as combinations of such forms. For example, instead of the function $f_\alpha(X)$, we may consider the function

$$y_\alpha = \bar{f}_\alpha(X) = \sum_{i=1}^{m_{\alpha 1}} b_i^{(\alpha 1)} x_1^{\alpha_{i1}^{(\alpha 1)}} x_2^{\alpha_{in}^{(\alpha 1)}} + \frac{\sum_{i=1}^{m_{\alpha 2}} b_i^{(\alpha 2)} x_1^{\alpha_{i2}^{(\alpha 1)}} x_2^{\alpha_{i2}^{(\alpha 2)}} \cdots x_n^{\alpha_{in}^{(\alpha 2)}}}{\sum_{i=1}^{m_{\alpha 3}} b_i^{(\alpha 3)} x_1^{\alpha_{i1}^{(\alpha 3)}} x_2^{\alpha_{i2}^{(\alpha 3)}} \cdots x_n^{\alpha_{in}^{(\alpha 3)}}} \quad (\alpha = 1,\ldots,k), \tag{7.5}$$

while instead of $g_\beta(X)$

$$\bar{g}_\beta(X) = \sum_{i=1}^{l_{\beta 1}} d_i^{(\beta 1)} x_1^{c_{i1}^{(\beta 1)}} x_2^{c_{i2}^{(\beta 1)}} \cdots x_n^{c_{in}^{(\beta 1)}} + \frac{\sum_{i=1}^{l_{\beta 2}} d_i^{(\beta 2)} x_1^{c_{i1}^{(\beta 2)}} x_2^{c_{i2}^{(\beta 2)}} \cdots x_n^{c_{in}^{(\beta 2)}}}{\sum_{i=1}^{l_{\beta 3}} d_i^{(\beta 3)} x_1^{c_{i1}^{(\beta 3)}} x_2^{c_{i2}^{(\beta 3)}} \cdots x_n^{c_{in}^{(\beta 3)}}} \quad (\beta = 1,\ldots,p). \tag{7.6}$$

In the formulas (7.5), (7.6), the exponents $\alpha_{ij}^{(\alpha\gamma)}$, $c_{ij}^{(\beta\gamma)}$ $(\gamma=1,\ldots,n;\ \gamma=1,2,3)$ may be any real numbers.

The inequalities (7.3) determine the admissible region Ω in the space of variables $x_1,\ldots,x_n$ for the variable parameters of the system. It is likely that the region Ω will not turn out to be convex.

For the solution of the stated vector optimization problem, we apply the method of approximation of the functions (7.5) to their ideal values in the space of system criteria.

In each scalar optimization problem, i.e. for a fixed function $y_\alpha(X)$ and constraints (7.3), (7.6), let there be determined the optimal vector

$$X_\alpha^0(x_1^{(\alpha)}, x_2^{(\alpha)}, \ldots, x_n^{(\alpha)}) \quad (\alpha=1,\ldots,k), \tag{7.7}$$

minimizing the functions y_α, respectively. We compute the optimal values of the functions $y_1, y_2, \ldots, y_k$ corresponding to the optimal vectors, i.e. we determine

$$y_\alpha^0 = \overline{f}_\alpha(X_\alpha^0) \quad (\alpha=1,\ldots,k). \tag{7.8}$$

The point y^0 with coordinates $\{y_1^0, y_2^0, \ldots, y_k^0\}$ in the k-dimensional space of the functions $y_1, y_2, \ldots, y_k$, is called the ideal or utopian point. Naturally, the components of the utopian point correspond to different vectors X_α^0, since the expression

$$x_i^{(1)} = x_i^{(2)} = \ldots = x_i^{(k)} \quad (i=1,\ldots,n)$$

is satisfied in few cases.

We form the approximation function

$$R(X) = \sum_{\alpha=1}^{n} \left(\frac{y_\alpha(X) - y_\alpha^0}{y_\alpha^0} \right)^2, \tag{7.9}$$

in which $y_\alpha(X)$ are expressed as in (7.5), while the numbers y_α^0 are determined according to (7.8).

Now we formulate the vector optimization problem: determine those vectors $X^*(x_1^*, x_2^*, \ldots, x_n^*)$, for which the function (7.9) is minimized and which satisfy the constraints

$$\begin{gathered} x_1 \geq 0, x_2 \geq 0, \ldots, x_n \geq 0, \\ \overline{g}_1(X) \leq 0, \overline{g}_2(X) \leq 0, \ldots, \overline{g}_p(X) \leq 0, \end{gathered} \tag{7.10}$$

where $\overline{g}_\beta(X)$ is determined according to (7.6). Such a vector $X^*(x_1^*, x_2^*, \ldots, x_n^*)$ is called a solution to the vector optimization problem.

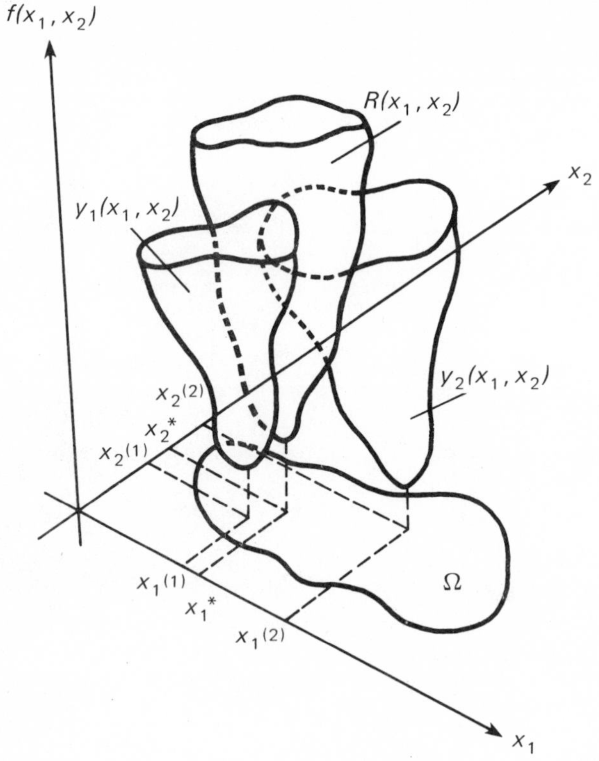

Fig. 13

The possible geometric locations of the functions $y_1(X)$, $y_2(X)$, $R(X)$, and the values of the vectors $X_1^0(x_1^{(1)}, x_2^{(2)})$, $X_2^0(x_1^{(2)}, x_2^{(2)})$, $X^*(x_1^*, x_2^*)$, for the case $n = 2$, in the space $\{x_1, x_2, f(x_1, x_2)\}$ is shown in Fig. 13.

An important application of the developed theory is in practical energy development projects. We apply the theory for calculation of optimal parameters for a heat exchanger for a nuclear power station.

2. The Problem of Parameter Optimization for a Heat Exchanger in a Nuclear Power Station

The predicted world energy supply shows that in the year 2000 half of all the electrical energy in the world will be supplied by nuclear power stations (AES). In correspondence with the world tendency and complicated by the rate of consumption of electrical energy, by 1990 in the USSR the capacity of nuclear power generation must be increased by 130-180 million kwh, while by the year 2000 it must be increased by 400-600 million kwh.

In connection with these requirements, the 1980-2000 production requirements for AES is very high, and basic new approaches to their construction are required. In particular, for the construction of heat exchange devices for AES of large capacity with fast neutron reactors, besides insuring normal functioning of the heat exchangers, it is necessary to minimize their weight, size, cost or some other important techno-economic aspect of the system. In this connection, it is appropriate to calculate heat exchanger parameters by formulating the situation as an optimization problem (for example, see (187)).

However, in a series of cases, it is necessary to optimize the parameters with respect to conflicting criteria, having different physical character. In such cases, it is advisable to reduce the choice of parameters to the solution of a vector optimization problem.

For example, for power stations of all types, there exist two criteria characterizing their economics: a) the capital expenditure for construction (the cost of a kilowatt). These costs are very high for hydro power stations, and somewhat less for nuclear and comparatively low for conventional power stations; b) the price (production cost) of electrical energy. This cost is cheaper for hydro power stations and costly for heating-type power stations. The second criterion depends upon the coefficient of efficiency of the operation cycle, the type of equipment, the cost of fuel, and so on. It is very important to note that the best type of power station for the first criterion is very bad relative to the second and conversly. For effective solution of such a problem, we resort to representing it as a vector optimization problem.

In the Institute of Nuclear Energy of the Academy of Sciences of the Byelorussian SSR, a single-circuit nuclear power station (AES) was designed with a 1 million kwh capacity, having a fast neutron reactor with a condensing dissociating heat exchanger--the BRGD-1000 (188).

The structure of the production cost of electrical energy shows that in nuclear stations with fast neutron reactors, the major fraction of the cost is that of the equipment (189). In turn, some of the most important pieces of equipment in an AES are the heat exchangers.

As characteristics show, the specific weight of heat exchangers in the total cost of an AES is very high (from 17% to 27.2% of the cost for the BRGD-1000). Thus, the investment in a nuclear station is, to a large degree, determined by the cost of heat exchangers. Consequently, optimization of heat exchange parameters from the point of view of minimization of their cost has great practical value.

It is necessary to note that the variation of several parameters in a heat exchanger also influences the characteristics of other equipment in the AES. So, for example, variation of the minimal temperature or the pressure in the regenerators, for the same heat capacity of a reactor, leads to a change in the electrical capacity of the station; such a parameter, since we have a multiplicity of coolers in the condenser, strongly influences the cost of the water supply for an AES.

An approach to the problem of designing a nuclear power station only from the point of view of minimizing equipment cost will not be of complete value. In different cases, important roles are played by the size and weight of the equipment. These requirements must be taken into account in the design of stationary, floating or movable AES.

3. The Parameter Optimization Problem for the Condenser of the BRGD-1000

We consider the problem of optimizing the parameters of the condensers in the AES BRGD-1000. Depending upon the operational situation in AES of the stated type, the condensers may have different, and sometimes conflicting requirements. In this connection, the solution of the compromise optimization problem by means of a complex scalar criterion is, in many cases, extremely difficult or sometimes impossible.

The theoretical machinery of vector optimization allows the engineer-designer to formulate in each case a vector criterion for evaluating choices of condenser parameters.

The positive aspect of such an approach, everything else being equal, is that in the passage from one vector optimization problem to another for a given type of condenser, variations which must be introduced into the computational formulas are negligible.

Analysis of the parameter optimization problem of the condenser in general show that, with regard to specific optimization problems, in particular for effective analytic representation of the objective function and convenient construction of an algorithm, it is advisable to identify four independent parameters of the condenser, which may have an affect on its size, weight and cost indices:

x_1 - the internal diameter of the condenser pipes;

x_2 - the velocity of the water in the condenser pipes;

x_3 - the cooling multiplier (the relation between the expense of cooling water to the expense of the heating medium);

x_4 - the initial temperature of the cooling water.

In many cases, it is necessary to calculate the parameters of a condenser with minimal weight and the minimal cost cooling system corresponding to it. Since these two requirements are contradictory, the choice of the best compromise solution, in general, must be sought using a vector criterion.

On the basis of the given data and the equations of heat transfer, we calculate the length and number of pipes. The thickness of the pipe walls is determined by technological considerations. On the basis of this information, taking account of construction considerations, we reduce the functional dependence of the weight of all condensers to the four above-mentioned parameters:

$$f_k(x) = 1.45\alpha_3 \frac{x_1 + 0.5\alpha_4}{x_1} f_e(x), \tag{7.11}$$

where the expression

$$f_e(x) = \frac{x_3}{2\alpha_7} \ln \frac{\alpha_5 - x_4}{\alpha_5 - x_4 - \dfrac{2\alpha_7}{x_3}} \left(\frac{1}{\alpha_b} + \frac{x_1}{\alpha_9} \ln \frac{x_1 + \alpha_4}{x_1} + \frac{x_1}{\alpha_k(x_1 + \alpha_4)}\right) \tag{7.12}$$

is introduced in the interest of simplifying the solution algorithm. The functions α_b and α_k entering into (7.12) determine, respectively, the coefficients of thermal efficiency of the water and the thermal medium and are calculated by the formulas

$$\alpha_b = \alpha_6 x_1^{-0.2} x_2^{0.8}, \tag{7.13}$$

$$\alpha_k = \alpha_8 (x_1 + \alpha_4)^{-0.25} (\alpha_5 - T^0)^{-0.25}. \tag{7.14}$$

The value T^0 is determined for each fixed pair x_3, x_4 by the method of successive approximations using the formula

$$T^0 = \frac{\alpha_k \alpha_5 + \alpha_b \left(x_4 + \dfrac{\alpha_7}{x_3}\right)}{\alpha_k + \alpha_b}. \tag{7.15}$$

The initial approximation for T^0 is taken to be

$$T^0 = 0.5\left(\alpha_5 + x_4 + \frac{\alpha_7}{x_3}\right). \tag{7.16}$$

For computation of the cost of the system's water supply, the total capital expenditure is divided into two components: the first depends only upon the expense of the cooling water (the cost of the pumping stations, underwater and above-water canals and so on), the second, the capital expenditure in the cooling process, itself, depends principally upon the attained temperature of the cooling water and partly upon its consumption. The second part of the cost is given in Table 8 by the function $f_2(x_3,x_4)$ for a region in the central European part of the USSR.

TABLE 8.

Capital Expenditure for Cooling Water--the Function $f_2(x_3,x_4)$ for a Region in the Central European Part of the USSR (in millions of rubles)

Temperature x_4 ^{0}C	Cooling Multiplicity							
	5	10	15	20	25	30	35	40
12.5	6.8	7.9	9.0	10.3	11.2	12.0	12.8	13.6
14.5	6.1	6.85	7.6	8.35	9.2	9.6	10.3	10.9
16.5	4.94	5.77	6.6	7.22	7.8	8.1	8.8	9.35
18.5	4.4	4.95	5.5	6.14	6.6	7.0	7.6	8.0
20.5	3.9	4.35	4.8	5.2	5.8	6.2	6.7	7.2

Remarks: 1. For computing the objective function, the numerical value of the function f_2 is substituted in rubles, i.e., the quantities presented in Table 8, are multiplied by 10^6.

2. Intermediate values of the function are obtained by linear interpolation.

In general form, for determination of the cost of the water cooling system we have the following rule

$$f_s(x) = \alpha_{10}x_3 + f_2(x_3,x_4). \tag{7.17}$$

TABLE 9.

Range of Variation of the Independent Variables

i	Dimension of x_i	x_i^{min}	x_i^{max}	Step Size
1	m	8.10^{-3}	30.10^{-3}	1.10^{-3}
2	m/sec	0.5	3.0	0.1
3	—	5	40	1
4	^{0}C	12.5	20.5	0.5

For optimization of the parameters, we impose the linear constraints

$$x_i^{min} \le x_i \le x_i^{max} \quad (i=1,2,3,4) \tag{7.18}$$

(cf. Table 9) and the nonlinear constraint

$$N = \alpha_{11}x_1^{-0.2}x_2^{2.8}f_e(x) \le N_0, \tag{7.19}$$

in the capacity of cooling water circulation, where N_0 is the maximal attainable circulation of cooling water.

The coefficients α_i (i=1,2,...,11) entering for convenience of representing (7.11) - (7.19), are constants for each different condenser variant and are considered to be given inputs.

Computational formulas for the coefficients α_i and for the input data are given in Tables 10 and 11, respectively.

Thus, we have the following optimization problem: in the region Ω of admissible values of the parameters x_1, x_2, x_3, x_4, determined by (7.18) and (7.19), find those vectors $x^*(x_1^*,\ldots,x_4^*)$, which minimize the functions (7.11) and (7.17) simultaneously.

According to section 1 of this chapter, the given problem reduces to the minimization of the expression

TABLE 10.

Formulas for Computing the Coefficients α_i

Index	Dimension of the Coefficient	Formula for Computation
1	rubles/kg	α_1 given as input data
2	-	α_2
3	$\frac{\text{kcal} \cdot \text{kg}}{\text{m}^2 \cdot \text{hr}}$	$\alpha_3 = 3600\psi_p\gamma_{mp}G\Delta h\delta$
4	m	$\alpha_4 = 2\delta$
5	^{0}C	$\alpha_5 = T_s$
6	$\frac{\text{kcal} \cdot \text{sec}^{0.8}}{\text{m}^{2.6} \cdot \text{hr} \cdot \text{grad}}$	$\alpha_6 = 0.023p_{rb}^{0.43}\lambda_b\nu_b^{-0.8}$
7	^{0}C	$\alpha_7 = \frac{\Delta h}{2c_{pb}}$
8	$\frac{\text{kcal}}{\text{hr} \cdot \text{grad}^{3/4} \cdot \text{m}^{7/4}}$	$\alpha_8 = 0.725\phi_p\eta_p\left[\frac{3600\lambda_k^3\gamma_k(\gamma_k-\gamma_k'')\Delta h}{\mu_k}\right]^{0.25}$
9	$\frac{\text{kcal}}{\text{m} \cdot \text{hr} \cdot \text{grad}}$	$\alpha_9 = 2\lambda_{mp}$
10	rubles	$\alpha_{10} = \frac{3600G \cdot c_b}{\gamma_b}$
11	$\frac{\text{kw} \cdot \text{sec}^{4.8} \cdot \text{kcal}}{\text{n}^{5/6} \cdot \text{hr}}$	$\alpha_{11} = \frac{0.803k_p G\Delta h\gamma_b(\nu_b)^{0.2}}{\eta \cdot g}$

$$R = \left(\frac{f_k(x)-f_k^0}{f_k^0}\right)^2 + \left(\frac{f_s(x)-f_s^0}{f_s^0}\right)^2, \tag{7.20}$$

in the region Ω, in which $f_k(x)$ and $f_s(x)$ are defined by formulas (7.11) and (7.17), respectively, while the numbers f_k^0 and f_s^0 are computed as

$$f_k^0 = \min_{x\in\Omega} f_k(x), \tag{7.21}$$

$$f_s^0 = \min_{x\in\Omega} f_s(x). \tag{7.22}$$

TABLE 11.
Table of Input Data

No.	Quantity	Notation	Dimension	Value for one of the Variations
1	2	3	4	5
1	consumption of heating medium	G	kg/sec	4800
2	dimension of enthalpy in the heat medium	Δh	kcal/kg	151
3	maximal capacity for circulation of cooling water	N_0	kw	10000
4	specific weight of the pipe material	γ_{mp}	kg/m^3	7850
5	specific weight of the heat medium in the liquid state	γ_k	kg/m^3	1396.1
6	specific weight of the heat medium in the gaseous state	γ''_k	kg/m^3	6.92
7	specific weight of water	γ_b	kg/m^3	998.2
8	Prandtl number of water	P_{rb}	-	7.0
9	heat capacity of water	c_{pb}	$\frac{kcal}{kg \cdot grad}$	0.999
10	coefficient of kinematic viscosity of water	ν_b	m^2/sec	$1.002 \cdot 10^{-6}$
11	coefficient of dynamic viscosity of the heat medium (condensers)	μ_k	$\frac{kg/sec}{m^2}$	$0.338 \cdot 10^{-4}$
12	thermal conductivity of the condensate	λ_k	$\frac{kcal}{m \cdot hr \cdot grad}$	0.1035
13	thermal conductivity of the pipe material	λ_{mp}	$\frac{kcal}{m \cdot hr \cdot grad}$	14.0

TABLE 11 (cont.)

1	2	3	4	5
14	thickness of the pipe walls	δ	m	10^{-3}
15	coefficient accounting for the weight of the cooling fins	ψ_p	--	1
16	coefficient of the cooling fins	ϕ_p	--	1
17	coefficient accounting for the effectiveness of heat transfer of the cooling fins	η_p	--	1
18	coefficient of pumping efficiency	η	--	0.65
19	gravitational strength	g	m/sec^2	9.81
20	coefficient	α_1	rubles/kg	5.45
21	coefficient	α_2	--	2.38
22	coefficient accounting for local loss of pressure	k_p	--	1.5
23	coefficient	e_1	rubles/kg	$-0.3186 \cdot 10^{+1}$
24	coefficient	e_2	rubles/kgm	$+0.9268 \cdot 10^{+2}$
25	coefficient	e_3	rubles/kg	$+0.9991 \cdot 10^{-1}$
26	coefficient	e_4	$rubles \cdot m^2/kg$	$-0.1838 \cdot 10^{-3}$
27	saturation temperature	T_2	^0C	36.5
28	thermal conductivity of water	λ_b	$\frac{kcal}{m \cdot hr \cdot grad}$	0.516
29	specific expenditures on hydrotechnical structures	c_b	$rubles \cdot hr/m^3$	15

The requirements that the condenser weight and the cost of cooling water be minimized are conflicting.

Introduction of the criterion (7.20) is logically justifiable. It is based upon the following argument: in view of the conflicting requirements, there does not exist an $\bar{x} \in \Omega$, for which we can simultaneously satsify the relations $f_k(\bar{x}) = f_k^0$, $f_s(\bar{x}) = f_s^0$. Therefore, it is advisable to assume that the best such point $x^* \in \Omega$, the representation of which $f^*(f_k(x^*), f_s(x^*))$ on $F = F_k \times F_s$ $(F_k = \{f_k(x) \mid x \in \Omega\};\ F_s = \{f_s(x) \mid x \in \Omega\})$, minimizes (7.20) under the condition that a euclidean metric is introduced into F. In other words, the optimal parameters x_1^*, x_2^*, x_3^*, x_4^* are sought to minimize the distance between the vectors $f^* = \{f_k(x^*), f_s(x^*)\}$ and $f^0 = \{f_k^0, f_s^0\}$ in the euclidean space of $\{f_k(x), f_s(x)\}$.

The given problem was solved for the condenser variation given by the input data in Tables 8-11. We obtained the following results. The minimal condenser weight was obtained for the parameter values.

$$x_1' = 0.008 \text{ m}; \qquad x_2' = 2.99 \text{ m/sec};$$
$$x_3' = 39; \qquad x_4' = 12.5^0 \text{ C} \tag{7.23}$$

and equals

$$f_k^0 = f_k(x_1', x_2', x_3', x_4') = 683{,}852 \text{ kg}. \tag{7.24}$$

Under the parameter values (7.23), the cost of the system's cooling water equals $f_s(x_1', x_2', x_3', x_4') = 23{,}615{,}924$ rubles.

The minimal cost of the system's cooling water is achieved for the parameter values

$$x_1'' = 0.0008 \text{ m}; \qquad x_2'' = 1.83 \text{ m/sec};$$
$$x''_3 = 10; \qquad x_4'' = 20.5^0 \text{ C} \tag{7.25}$$

and equals

$$f_s^0 = f_s(x_1'', x_2'', x_3'', x_4'') = 6{,}946{,}674 \text{ rubles.} \qquad (7.26)$$

Under the parameter values (7.25), the weight of the condensers equals

$$f_k(x_1'', x_2'', x_3'', x_4'') = 2{,}697{,}602 \text{ kg.}$$

For the solution of the vector optimization problem, we minimize (7.20) and the parameters are determined as

$$x_1^* = 0.008 \text{ m}; \quad x_2^* = 2.27 \text{ m/sec};$$
$$x_3^* = 11; \quad x_4^* = 19.1^0\text{C.} \qquad (7.27)$$

The obtained parameters (7.27) yield the optimal values of (7.11) and (7.17) as

$$f_k^* = f_k(x_1^*, x_2^*, x_3^*, x_4^*) = 1{,}491{,}846 \text{ kg,}$$
$$f_s^* = f_s(x_1^*, x_2^*, x_3^*, x_4^*) = 7{,}945{,}244 \text{ rubles.} \qquad (7.28)$$

Consequently, the condenser design corresponding to the parameters (7.27), has the following compromise features: for an increase in the capital cost of one million rubles in construction of the cooling system, the weight of the condensers may be reduced from 2700 tons to 1500 tons. Or, increasing the weight of condensers by 800 tons, the cooling system can be constructed for 15.6 million rubles less.

It is of interest to consider the solution of the concrete problem of optimizing condenser parameters from the point of view of a complex criterion presented by engineers at the Institute for Nuclear Energy of the BSSR Academy of Sciences. On the basis of substantial experience in the construction of heat exchangers, as well as from physical considerations, the solution of the particular design problem requires us to determine optimal condenser parameters, corresponding to minimizing the function

$$f(x) = \lambda_1 f_k(x) + \lambda_2 f_s(x), \tag{7.29}$$

in which the coefficients λ_1, λ_2 are determined as

$$\lambda_1 = \frac{\alpha_1 + \alpha_2 f_1(x_1)}{1.45}, \qquad \lambda_2 = 1. \tag{7.30}$$

Here the function $f_1(x_1)$ determines the cost of 1 kg of smooth or finned pipe and is representable in the approximate form

$$f_1(x_1) = e_1 + e_2 x_1 + e_3 x_1^{-1} + e_4 x_1^{-2}, \tag{7.31}$$

where the coefficients e_1, e_2, e_3, e_4 are given as input data, together with a_1 and a_2 in Table 11.

Minimization of (7.29) is carried out under the same constraints (7.18), (7.19). Carrying out the computations determines the following parameters for the condensers:

$$\begin{aligned} \tilde{x}_1 &= 0.019 \text{ m}; & \tilde{x}_2 &= 2.58 \text{ m/sec}; \\ \tilde{x}_3 &= 10; & \tilde{x}_4 &= 12.5^0\text{C}, \end{aligned} \tag{7.32}$$

which, in turn determine the weight of the condensers and the cost of the cooling water as

$$\begin{aligned} \tilde{f}_k &= f_k(\tilde{x}_1, \tilde{x}_2, \tilde{x}_3, \tilde{x}_4) = 1{,}141{,}164 \text{ kg}, \\ \tilde{f}_s &= f_s(\tilde{x}_1, \tilde{x}_2, \tilde{x}_3, \tilde{x}_4) = 10{,}515{,}224 \text{ rubles}. \end{aligned} \tag{7.33}$$

Comparison of the results (7.32) and (7.33) with the results (7.27) and (7.28), their closeness clearly testifies to the effectiveness of the vector approach to the problem of optimal system design under deterministic conditions, i.e. when we have sufficient information about the coefficients λ_1 and λ_2.

During the design of the condensers of a nuclear power station, different requirements are presented. Such requirements may be both minimization of the volume of the condensers and minimization of the overall cost of the condensers, together with the water cooling system. Now we solve such a problem for the condenser variation of the same type as in the BRGD-1000.

The expression for the functional dependence of the volume of the condenser upon the parameters x_1, x_2, x_3, x_4 has the form

$$f_{ob} = 3214G\Delta h \frac{(x_1+\alpha_4)^2}{x_1} f_e(x), \tag{7.34}$$

while the total cost of the condensers with the water cooling system is expressed as

$$f_{st} = (\alpha_1+\alpha_2 f_1(x_1)) \frac{\alpha_3(x_1+0.5\alpha_4)}{x_1} f_e(x) + \alpha_{10}x_3 + f_2(x_3,x_4). \tag{7.35}$$

The coefficients and functions entering into (7.34) and (7.35) have the same meaning as above and are defined in Tables 8-11. Actually, (7.35) is the same as (7.29) under the conditions (7.11), (7.17), (7.30).

The optimization problem consists of the following: in the region Ω of admissible values of the parameters x_1, x_2, x_3, x_4, determined by (7.18) and (7.19), find those vectors $X^*(x_1^*,x_2^*,x_3^*,x_4^*)$ for which the functions (7.34) and (7.35) assume their minimum values.

For solution of this problem, we obtain the following results.

The minimal condenser size is assumed for the parameter values

$$x_1' = 0.008 \text{ m}; \quad x_2' = 2.99 \text{ m/sec};$$
$$x_3' = 39; \quad x_4' = 12.5^0\text{C} \tag{7.36}$$

and equals

$$f_{ob}^0 = f_{Ob}^0(x_1', x_2', x_3', x_4') = 596.0 \text{ m}^3. \tag{7.37}$$

For the values (7.36), the overall cost of the condensers and cooling system equals $f_{st}(x_1', x_2', x_3', x_4') = 34{,}090{,}748$ rubles.

The minimal total cost is achieved for the parameter values

$$x_1'' = 0.019 \text{ m}; \quad x_2'' = 2.58 \text{ m/sec};$$
$$x_3'' = 10, \quad x_4'' = 12.5^0\text{C} \tag{7.38}$$

and equals

$$f_{st}^0 = f_{st}(x_1'', x_2'', x_3'', x_4'') = 21{,}011{,}898 \text{ rubles}. \tag{7.39}$$

For the parameter values (7.38), the condenser size is $f_{ob}(x_1'', x_2'', x_3'', x_4'') = 1909.5 \text{ m}^3$.

The compromise solution of the problem, based on minimization of the measure of approximation

$$R = \left(\frac{f_{ob}(x) - f_{ob}^0}{f_{ob}^0}\right)^2 + \left(\frac{f_{st}(x) - f_{st}^0}{f_{st}^0}\right)^2, \tag{7.40}$$

is determined by the following parameter values

$$x_1^* = 0.008 \text{ m}; \quad x_2^* = 2.71 \text{ m/sec};$$
$$x_3^* = 17; \quad x_4^* = 15.6^0\text{C}. \tag{7.41}$$

The parameters (7.41) yield the following values for the functions (7.34) and (7.35):

$$f^*_{ob} = f_{ob}(x_1^*, x_2^*, x_3^*, x_4^*) = 785.6 \text{ m}^3$$
$$f^*_{st} = f_{st}(x_1^*, x_2^*, x_3^*, x_4^*) = 25{,}855{,}837 \text{ rubles.} \qquad (7.42)$$

Consequently, design of the condenser, satisfying the parameters (7.41), has the following compromise aspect: for an increase in the overall cost of the condensers and water cooling system of 4.8 million rubles, we can decrease the size of the condensers from 1.9 thousand cubic meters to 786 cubic meters. Or, for an increase of 190 cubic meters in condense size, we can decrease the overall cost by 8.2 million rubles.

For creation of a more perfect condenser design, attention must be paid to requirements of weight and cost, as well as the volume, simultaneously.

In spite of the fact that the minimal volume of the condensers is assumed for the parameters (7.23), which simultaneously minimize the condenser weight, minimization of (7.20) does not consider the requirement on the size index of the condenser.

In such a case, we are led to minimization of the expression

$$R(x) = \left(\frac{f_k(x) - f_k^0}{f_k^0}\right)^2 + \left(\frac{f_s(x) - f_s^0}{f_s^0}\right)^2 + \left(\frac{f_{ob}(x) - f_{ob}^0}{f_{ob}^0}\right)^2, \qquad (7.43)$$

in which the functions $f_k(x)$, $f_s(x)$, $f_{ob}(x)$ are naturally determined in the forms (7.11), (7.17), (7.34), while the numbers f_k^0, f_s^0, f_{ob}^0 equal the values (7.24), (7.26), (7.37), respectively.

Minimization of the function R(x), defined by (7.43), determines the parameters of the condenser in Ω as

$$\bar{x}_1^* = 0.008\ \text{m}; \qquad \bar{x}_2^* = 2.69\ \text{m/sec};$$

$$\bar{x}_4^* = 10; \qquad \bar{x}_4^* = 12.5^{\circ}\text{C}, \tag{7.44}$$

which are optimal in the sense of the three stated performance criteria: weight, cost and size.

For the parameters (7.44), the stated criteria assume the values

$$\bar{f}_k^* = f_k(\bar{x}_1^*, \bar{x}_2^*, \bar{x}_3^*, \bar{x}_4^*) = 926{,}175\ \text{kg};$$

$$\bar{f}_s^* = f_s(\bar{x}_1^*, \bar{x}_2^*, \bar{x}_3^*, \bar{x}_4^*) = 10{,}762{,}712\ \text{rubles}; \tag{7.45}$$

$$\bar{f}_{ob}^* = f_{ob}(\bar{x}_1^*, \bar{x}_2^*, \bar{x}_3^*, \bar{x}_4^*) = 807.0\ \text{m}^3.$$

For clarity, we summarize our results in Table 12.

TABLE 12.

x_1 [m]	x_2 [m/sec]	x_3 —	x_4 [^{0}C]	Condenser Weight f_k [kg]	Cost of Cooling System f_s [rubles]	Condenser Size f_{ob} [m^3]
0.008	2.99	39	12.5	$f_k^0 = 683{,}852$	23,615,924	$f_{ob}^0 = 596.0$
0.008	1.83	10	20.5	$f_k = 2{,}697{,}602$	$f_s^0 = 6{,}946{,}674$	2352.0
0.008	2.27	11	19.1	$f_k^* = 1{,}491{,}846$	$f_s^* = 7{,}945{,}244$	1300.7
0.019	2.58	10	12.5	$\tilde{f}_k = 1{,}141{,}164$	$\tilde{f}_s = 10{,}515{,}224$	1909.5
0.008	2.69	10	12.5	$\bar{f}_k^* = 926{,}175$	$\bar{f}_s^* = 10{,}762{,}712$	$\bar{f}_{ob}^* = 807.0$

The results (7.44), (7.45) coincide completely with the requirements which are presented in the design of condensers in nuclear power stations.

The calculated optimal variants of condensers correspond to the following parameters: the condensers must be constructed with pipes having an interior diameter of 8mm; the velocity of the protective water in the condenser pipes must be 2.69 m/sec; the multiplicity factor of cooling equals 10; the initial temperature of the water must be 12.5°C.

A condenser constructed according to these parameters, as compared with a condenser with the parameters (7.32), will have a weight of 215 tons less and will have a volume of 1.2 thousand cubic meters less. This will be realized with an increase of capital expenditure for the system water cooling structure of only 247.5 thousand rubles.

4. A Search Algorithm

The solution of each minimization problem at all steps is carried out by a search method. Toward this end, we use a search procedure of gradient-type, developed in the Control Systems Institute of the GSSR, which takes into account locations of extrema in similar sorts of problems and which is the best modification of a local method presented in (187). Since it is not certain that the function has a single extremum, the search is conducted at each step from a multitude of initial points. Then we choose the smallest of the obtained minima.

The specific character of the determination of the extrema in such types of problems is that the solution, as a rule, is found on a nonlinear boundary. Therefore, the work of the algorithm is organized so that it is somewhat quicker to go out to the boundary and to move along it in the direction of increase of the objective function.

A general scheme of motion along the boundary is the following. From the boundary, we make a step along the antigradient of the objective function $Q(x)$, as a rule in a prohibited region, until the function $Q(x)$ stops increasing. Next, we make a step along the antigradient of the constraint function $\phi(x)$ up to the boundary $d = \phi(x) - c = 0$. Thus, the operations of a return to the boundary and a step along it are combined.

The problem is formulated in the following form:

$$\{\min Q(x), \quad x: \ d = \phi(x) - c \le 0,$$

$$\alpha_\beta \le x_\beta \le b_\beta, \quad \beta = 1,\ldots,2,m\}.$$

The search is started from points in the interior or on the boundary of the admissible region. We make the next approximation by the following recurrence formula:

$$x^{(i+1,j)} = x^{(i+1,j-1)} - \lambda_j p^{(i+1)}$$

$$(i=1,2,\ldots;\ j=1,2,\ldots,n_{i+1}),$$

where i, j are integers of the stage and step, respectively. The vector $p^{(i+1)}$ determines the direction of movement.

For $d > \varepsilon$, where $\varepsilon > 0$ is a quantity given in advance for the admissible error for hitting the boundary ($\varepsilon=10^{-3}$), i.e. for movement in the interior of the admissible domain, the vectors $p^{(i+1)}(p_1^{(i+1)},\ldots,p_m^{(i+1)})$ are selected by the expression

$$p^{(i+1)} = \alpha_1 \operatorname{grad} Q(x^{(i,n_i)}) + \alpha_2 p^{(i)} \quad (i=1,2,\ldots).$$

Here $p^{(1)} = \operatorname{grad} Q(x^{(0,0)})$ is the gradient at the initial search point.

The choice of the search direction, i.e. the determination of the coefficients α_1, α_2 is helped to some degree by avoiding a zig-zag movement along the ravine. In our algorithm, they are defined to be $\alpha_1 = 0.5$, $\alpha_2 = 0.4$.

Choice of the step size in such situations is made in the following form

$$\lambda_j = \alpha_3 \lambda_{j-1} \qquad (j=2,3,\ldots,n_{i+1}),$$

where

$$\lambda_1 = \frac{\alpha_0}{|p^{(i+1)}|} .$$

The coefficients α_0, α_3 are chosen from practical considerations. For example, they are taken to be relatively large if it is necessary to quickly get to the boundary. On the other hand, when the objective function has many extrema and we are interested in local extrema points, these coefficients are taken to be small.

For motion along the boundary $|d| \le \varepsilon$, it is advisable to make steps in the direction of the gradient

$$p^{(i+1)} = \text{grad } Q(x^{(i,n_i)}).$$

The multiplier λ_i is computed in this case by the formula

$$\lambda_j = \left|\{\text{sign}\left[\text{grad } Q(x^{(i,n_i)})^T (x^{(i,n_i)} - x^{(i-1,n_i-1)})\right] - 0.5\}\lambda_{j-1}\right|.$$

Such a choice of λ_i allows us to reduce the step size from the previous step or to increase it, when the corresponding point passes to the opposite slope of the "ravine" or, conversely, moves along the same slope.

If the step results in an increase of the function, i.e.

$$\Delta Q^{(i+1,j)} = Q(x^{(i+1,j)}) - Q(x^{(i+1,j-1)}) > 0,$$

the multiplier λ_i is decreased further, relative to the last step, in order that the function at the last point be less than at the preceding point. The multiplier is chosen by the following rule:

$$\{\max_k \lambda,\ \lambda = \lambda_j \cdot 2^{-k}, Q(x^{(i+1,j)}+\lambda p^{(i+1)}) \leq Q(x^{(i+1,j)})\}.$$

If we hit points outside the region, i.e. when $d > \varepsilon$, we return to boundary points by means of

$$p^{(i+1)} = \text{grad}\ \phi(x^{(i,n_i')}),$$

where n_i' is the index corresponding to the point hitting outside the region.

The multiplier λ_j is chosen by the rule

$$\lambda_j = \alpha_4 \frac{d^{(j)}}{|\Delta d^{(j)}|} \lambda_{j-1} \quad (j=1,2,\ldots,n_{i+1}), \tag{7.46}$$

where

$$\Delta d^{(j)} = \phi(x^{(i+1,j)}) - \phi(x^{(i+1,j-1)})$$

is the increment of the constraint function at the first step, where

$$\lambda_1 = \alpha_4 \frac{d^{(1)}}{\left|\text{grad}\ \phi(x^{(i,n_i')})\right|^2}. \tag{7.47}$$

In our algorithm $\alpha_4 = 1.2$.

Since the constraint functions may also turn out to be non-convex, for movement toward the boundary we could have the case $\Delta d^{(j)} \geq 0$. Here it is necessary to change the direction of ascent, correcting the previous step by the rule

$$\{\max_k \lambda,\ \lambda = \lambda_j \cdot 2^{-k}, \phi(x^{(i+1,j)}+\lambda p^{(i+1)}) \leq \phi(x^{(i+1,j)})\}.$$

The new direction is chosen by the rule

$$p^{(i+1)} = \text{grad}\ \phi(x^{(i+1,j)}).$$

We then proceed with a movement using a step chosen by (7.46) and (7.47).

We note that movement according to the rules (7.46), (7.47) is equivalent to the modification of Newton's method for finding the root of the equation

$$d(\lambda) = \phi(x^{(i,n_i)} - \lambda p^{(i+1)}) - c = 0,$$

in which the derivative is replaced by a finite difference.

Under movement from the inadmissible region to the boundary, in view of the non-convex constraint function it is possible to obtain a point in the admissible region. Here it is most advisable to move the point to the boundary. The corrective step in this case is carried out by the rule

$$\lambda_j = \frac{1}{1 - \dfrac{d^{(j)}}{d^{(j-1)}}} \lambda_{j-1}.$$

The search ends when the last step is less than a given quantity ε_β, i.e.

$$|x_\beta^{(i+1,n_{i+1})} - x_\beta^{(i,n_i)}| \leq \varepsilon_\beta \qquad \beta = 1,2,\ldots,m.$$

The described algorithm allows us to solve parameter optimization problems for system design in an acceptable amount of time. In particular, the solution of the parameter optimization problem for the condensers in the AES BRGD-1000 for ten initial search points, using an ALGOL program, was carried out in 1 hour on the M-22 computer.

BIBLIOGRAPHY

(Translator's Notes: 1) Many of the items cited below are Russian translations of books or articles originally published in English. Those publications of this type of which we are aware are marked with an asterisk (*). Also, many Russian journals are regularly translated into English, so the English reader interested in consulting any of the references should first check to see if the Russian original is available in English translation.

2) The following journal abbreviations are employed in the bibliography below:

(i) ABT = Abtomatika i Telemechanika (Automation and Remote Control)

(ii) PMM = Prikladnaya Matematika i Mechanika (Applied Mathematics and Mechanics).

1. Andronov, A.A., and Voznesenskii, I.N., Readings and Commentary in D.K. Maxwell, I.A. Vishnegradskii, A. Stodola, "The Theory of Automatic Control (Linearization Problems, Izd. Acad. Sci. USSR., ser. "Classical Science." (1949).
2. Petrov, B.N., Popov, E.P., et al., "Development of Control Theory in the USSR." Proc. 2nd All-Union Symp. on Control Theory, Vol. 1, Izd. Acad. Sci. USSR, (1955).
3. Mechanics in the USSR over the Last 50 Years, Vol. 1, General and Applied Mathematics, Izd. Nauka, (1968); ABT, Vol. 25, No. 6, (1964). 25th Anniversary Proc. of the Institute for Control Sciences; Soviet Control Science on the 50th Anniversary of the USSR, ABT, No. 11, (1972).
4. Ochotsmskii, D.E., The Theory of Rocket Motion, PMM 10, (1946).
5. Fel'dbaum, A.A., Optimal Processes in Automatic Control Systems, ABT 6, (1953).

*6. Pontryagin, L.s., et al., "The Mathematical Theory of Optimal Processes." Fizmatigiz, (1961).

*7. Boltyanskii, V.G., "Mathematical Methods of Optimal Control." Izd. Nauka, (1969).

*8. Bellman, R., "Dynamic Programming." Izd. Inostr. Litri, (1960).

*9. Bellman, R., and Dreyfus, S., "Applied Dynamic Programming." Izd. Nauka, (1965).
10. Krasovskii, N.N., "The Theory of Controlled Motion." Izd. Nauka, (1968).
11. Krasovskii, N.N., "Game Problems in Pursuit-Evasion Motion." Izd. Nauka, (1970).
*12. Isaacs, R., "Differential Games." Izd. Mir, (1967).
13. Rozonoer, L.I., The Pontryagin Maximum Principle in the Theory of Optimal Systems, I-II, ABT, Nos. 10, 11, 12, (1959).
*14. Gabasov, R., and Kirillova, F.M., "The Qualitative Theory of Optimal Processes." Izd. Nauka, (1971).
15. Gabasov, R., and Kirillova, F.M., "Singular Optimal Control." Izd. Nauka, (1973).
16. Kulikowski, R., "Optimal and Adaptive Processes in Automatic Control Systems." Izd, Nauka, (1967).
17. Krotov, V.F., and Gurman, V.I., "Methods and Problems of Optimal Control." Nauka, (1969).
18. Letov, A., "The Dynamics of Flight and Optimal Control." Nauka, (1969).
*19. Von Neumann, J., and Morgenstern, O., "The Theory of Games and Economic Behavior." Nauka, (1970).
*20. Luce, R., and Raiffa, H., "Games and Decisions." Izd, IL., (1961).
*21. Karlin, S., "Mathematical Methods in Game Theory, Programming and Economics." Mir, (1964)
22. Germeier, Yu.B., "Introduction to Operations Research." Nauka, (1971).
*23. Ventzel, E.S., "Operations Research." Izd. Soviet Radio, (1972).
*24. Hadley, G., "Nonlinear and Dynamic Programming." Mir, (1967).
25. Yudin, D.B., and Goldstein, E.G., "Linear Programming." Nauka, (1969).
*26. Duffin, R., E. Peterson, and Zener, C., "Geometric Programming." Mir, (1972).
*27. Dantzig, G., "Linear Programming." Progress, (1966).
28. Gabasov, R., and Kirillova, F.M., Application of the Maximum Principle for Computation of Optimal Control for Discrete Systems, "Doklady Acad. Sci. USSR, Vol. 189, No. 5, (1969).
29. Boltyanskii, V.G. "Optimal Control of Discrete Systems." Nauka, (1973).
*30. Tsypkin, Ya.Z., "Adaptation and Learning in Automatic Systems." Nauka, (1968).
31. Katkovnik, V.Ya., and Poluektov, R.A., "Multidimensional Discrete Control Systems." Nauka, (1966).
32. Propoi, A.I., "Elements of the Theory of Optimal Discrete Processes." Nauka, (1973).
33. Moiseev, N.N., "Numerical Methods in the Theory of Optimal Systems." Nauka, (1971).
34. Litovchenko, I.A., The Theory of Optimal Systems, in "All the Sciences, Mathematical Analysis, Probability Theory Control." Acad. Sci. USSR, (1962).

35. Krasovskii, N.N., Optimal Control in Ordinary Dynamical Systems, Uspekhi, Math. Nauk, Vol. 20, No. 3, (1965).
36. Krasovskii, N.N., and Moiseev, N.N., The Theory of Optimal Control Systems, TK, No. 5, (1967).
37. Letov, A.M., Optimal Control Theory (A Survey), in "Optimal Systems, Statisical Methods." Proc. 2nd IFAC Congress, Nauka, (1965).
38. Letov, A.M., Scientific Problems in Automatic Control, ABT, No. 8, (1966).
39. Letov, A.M., Some Unsolved Problems in Control Theory, Diff. Eqs., Vol. 6, No. 4, (1970).
40. Krasovskii, N.N, Gavrilov, M.A., Letov, A.M., and Pugachev, V.S., General Control Problems, Vestnik Acad. Sci. USSR, No. 8, (1970).
41. Kantorovich, L.V., Optimization Methods and Mathematical Models in Economics, Uspekhi Math. Nauk, Vol. 25, No. 5, (1970).
42. Krasovskii, N.N., The Theory of Optimal Control Systems, in "Mechanics in the USSR over the Last 50 Years." Vol. 1, (1968).
43. Gabasov, R., and Kirillova, F.M., The Contemporary State of the Theory of Optimal Processes, ABT, No. 9, (1972).
44. Palewonsky, B., Optimal Control: A Review of Theory and Practice, AIAA Journal, Vo. 3, No. 11, (1965).
45. Athans, M., The Status of Optimal Control Theory and Applications for Deterministic Systems, IEEE Trans. on Auto. Control, Vol. 11, No. 3, (1966).
46. Oldenburger, R., "Optimal Control." Holt, Rinehart and Winston, New York, (1966).
47. Flugge-Lotz, I., and Marbach, H., The Optimal Control of Some Attitude Control Systems for Different Performance Criteria, J. Basic Engrn., ASME Trns., Series D, Vol. 85, No. 2, (1963).
48. Nelson, W.L., On the Use of Optimization Theory for Practical Control System Design, IEEE Trans. on Auto. Control, Vol. AC-9, No. 4, (1964).
49. Pareto, V., "Cours d'Economie Politique." Lausanne, Rouge, (1896).
50. Debreu, G., "Theory of Value." John Wiley, New York, (1959).
51. Kuhn, H.W., and Tucker, A.W., Nonlinear Programming, in Proc. of the Second Berkeley Sym. on Mathematical Statistics and Probability, University of California Press, Berkeley, California, (1951).
52. Zadeh, L.A., Optimality and Non-Scalar-Valued Performance Criteria, IEEE Trans. on Auto. Control, Vol. AC-8, No. 1, (1963).
53. Klinger, A., Vector-Valued Performance Criteria, IEEE Trans. on Auto. Control, Vol. AC-9, No. 1, (1964).
54. Chang, Sheldon S.L., General Theory of Optimal Process, J. SIAM Control, Vol. 4, No. 1, (1966).

55. Da Cunha, N.O., and Polak, E., Constrained Minimization under Vector-Valued Criteria in Finite Dimensional Spaces, Electronics Research Laboratory, University of California, Berkeley, California, Memorandum No. ERL-188, (October, 1966).
56. Canon, M., Cullum, C., and Polak, E., Constrained Minimization Problems in Finite Dimensional Spaces, J. SIAM Control, Vol. 4, No. 3, (1966).
57. Da Cunha, N.O., and Polak, E., Constrained Minimization under Vector-Valued Criteria in Linear Topological Spaces, Electronics Research Laboratory, University of California, Berkeley, California, Memorandum No. ERL-191, (November, (1966).
58. Da Cunha, N.O., and Polak, E., Constrained Minimization under Vector-Valued Criteria in Linear Topological Spaces, in "Mathematical Theory of Control" (A.V. Balakrishnan and L.W. Neustadt, ed.). AP, New York and London, (1967).
59. Dong Hak Chyung, Optimal Systems with Multiple Cost Functionals, J. SIAM Control, Vol. 5, No. 3, (1967).
60. Waltz, F.M., An Engineering Approach: Hierarchical Optimization Criteria, IEEE Trans. on Auto. Control, Vol. AC-12, No. 2, (1967).
61. Reid, R., and Citron, S.J., On Noninferior Performance Index Vectors, J. of Optim. Theory and Appl., Vol. 7, No. 1, (1971).
62. Athans, M., and Geering, Hans, P., Necessary and Sufficient Conditions for Differentiable Nonscalar-Valued Functions to Attain Extrema, IEEE Trans. on Auto. Control, Vol. AC-18, No. 2, (1973).
63. Geering, H.P., "Optimal Control Theory for Non-Scalar-Valued Performance Criteria," Ph.D. Dissertation, Dept. Elect. Eng., M.I.T., Cambridge, (1971).
64. Kai-Ching, Chu, On the Noninferior Set for the Systems with Vector-Valued Objective Function, IEEE Trans. on Auto. Control, Vol. AC-15, No. 5, (1970).
65. Chattopadhyay, R., Linear Programming with Vector-Valued Cost Function, The Aloha System Report A 70-3, University of Hawaii, Honolulu, (1970).
66. Chattopadhyay, R., Well-Posed Linear Programs with Vector-Valued Cost Functions, Part 1, 2, The Aloha System Reports A 70-4 and A 70-5, University of Hawaii, Honolulu, (1970).
67. Chattopadhyay, R., Dominant Strategies in Linear Programs with Vector-Valued Cost Functions, The Aloha System Report A 70-7, University of Hawaii, Honolulu, (1970).
68. Chattopadhyay, R., Dominant Strategies and Super-Linear Programs with Vector-Valued Cost Functions, The Aloha System Report A 70-8, University of Hawaii, Honolulu, (1970).
69. Fishburn, P.C., A Study of Independence in Multivariate Utility Theory, Econometrica, Vol. 37, No. 1., (1969).
70. Benayoun, R., and Tergny, J., Critéres multiples et programmation mathematique: une solution dans le cas lineare, R.I.R.O., No. V-2, (1970).
71. Roy, B., Classément et choix en presence de points de vue multiples, R.I.R.O., No. 8, (1968).

72. Vincent, T.L., and Leitmann, G., Control Space Properties of Cooperative Games, J. of Optim. Theory and Appl., Vol. 6, No. 2, (1970).
73. Leitmann, G., and Stalford, H., A Sufficiency Theorem for Optimal Control, J. of Optim. Theory and Appl., Vol 8, No. 3, (1971).
74. Leitmann, G., Rocklin, S., and Vincent, T.L., A Note on Control Space Properties of Cooperative Games, J. of Optim. Theory and Appl., Vol. 9, No. 6, (1972).
75. Leitmann, G., and Schmitendorf, W., Some Sufficiency Conditions for Pareto-Optimal Control, Joint Automatic Control Conference, Stanford, (1972), to appear in J. of Dyn. Sys. Meas. and Control.
76. Blaquiere, A., Juricek, L, and Wiese, K.E., Geometry of Pareto Equilibria and a Maximum Principle in N-Person Differential Games, J. of Math. Anal. and Appl., Vol 38, Nol. 1, (1972).
77. Schmitendorf, W., Cooperative Games and Vector-Valued Criteria Problems, IEEE Trans. on Auto. Control, to appear (1973).
78. Stalford, H.L., Criteria for Pareto-Optimality, J. of Optim. Theory and Appl., Vol. 9, No. 6, (1972).
79. Ho, Differential Games, Dynamic Optimization and Generalized Control Theory, J. of Optim. Theory and Appl., Vol. 6, No. 3, (1970).
80. Stadler, W., A Note on the Insecurity of Nash-Equilibria, J. of Optim. Theory and Appl. (to appear).
81. Schmitendorf, W., Cooperative Games and Vector-Valued Criteria Problems, IEEE Trans. on Auto. Control, Vol. AC-18, No. 2, (1973).
82. Rekasius, Z.V., and Schmitendorf, W.E., On the Noninferiority of Nash-Equilibrium Solutions, IEEE Trans. on Auto. Control, Vol. AC-16, (1971).
83. Rekasius, Z.V., and Schmitendorf, W.E., Comments on On the Noninferiority of Nash-Equilibrium Solutions, IEEE Trans. on Auto. Control, Vol. AC-17, (1972).
84. Haurie, Alain, On Pareto Optimal Decisions for a Coalition of a Subset of Players, IEEE Trans. on Auto. Control, Vol. AC-18m No. 2, (1973).
85. Haurie, A., and Delfour, M., Individual and Collective Rationality in a Dynamic Pareto Equilibrium, J. of Optim. Theory and Appl. (to appear).
86. Blaquiere, A., Sur la géométrie des surfaces de Paréto d'un jeu différential à N jouers, C.R. Acad. Sci. Raris, sre. A, Vol. 271, (1970).
87. Blaquiere,A., Juricek, L., and Wiese, K.E., Sur la géométrie des surfaces de Paréto d'un jeu différential à N jouers; Théorème du maximum. C.R. Acad. Sci. Paris, Ser. A., Vol. 271, 1970.
88. Salama, A.I.A., "Optimization of Systems with Several Cost Functionals." Ph.D. Dissertation, Dept. of Elec. Eng., University of Alberta, Edmonton, Canada, (1970).

89. Salama, A.I.A., and Gourishankar, V., Optimal Control of Systems with a Single Control and Several Cost Functionals, Int. J. Control, Vol. 14, No. 4, (1971).

90. Salama, A.I.A., and Hamza, M.H., On the Optimization of Static Systems with Several Cost Measures, IEEE Trans. on Auto. Control, Vol. AC-17, No. 1, (1972).

91. Yu, R.L., A Class of Solutions for Group Decision Problems, Center for System Science 71-06, Graduate School of Management, University of Rochester, Rochester, New York, (1972).

92. Yu, P.L., Introduction to Domination Structures in Multicriteria Decision Problems, Systems Analysis Program, Working Paper Series No. F-7219, Graduate School of Management, University of Rochester, Rochester, New York, (1972).

93. Freimer, M., and Yu, P.L., The Application of Compromise Solutions to Reporting Games, Systems Analysis Program, Working Paper Series No. F-7220, Graduate School of Management, University of Rochester, Rochester, New York, (1972).

94. Yu, P.L., Nondominate Investment Policies in Stock Markets (Including and Empirical Study), Systems Analysis Program, Working Paper Series No. F-7222, Graduate School of Management, University of Rochester, Rochester, New York, (1972).

95. Yu, P.L., Cone Convexity, Cone Extreme Points and Nondominated Solutions in Decision Problems with Multiobjectives, Center for System Science 72-02, Graduate School of Management, University of Rochester, Rochester, New York, (1972).

96. Freimer, M., and Yu, P.L., An Approach toward Decision Problems with Multiobjectives, Center for System Science 72-03, Graduate School of Management, University of Rochester, Rochester, New York, (1972).

97. Yu, P.L., and Zeleny, M., The Set of all Nondominated Solutions in the Linear Cases and a Multicriteria Simplex Method, Center for System Science 73-03, Graduate School of Management, University of Rochester, Rochester, New York, (1973).

98. Yu, P.L., and Zeleny, M., On Some Linear Multi-Parametric Programs, Center for System Science 73-05, Graduate School of Management, University of Rochester, Rochester, New York, (1973).

99. Yu, P.L., and Leitmann, G., Compromise Solutions, Domination Structures, and Salukvadze's Solution, Center for System Science 73-08, Graduate School of Management, University of Rochester, Rochester, New York, (1973).

100. Bellman, R.E., and Zadeh, L.A., Decision-making in a Fuzzy Environment, Management Science, Vol. 17, No. 4, (1970).

101. Huang, S.C., Note on the Mean-Square Strategy for Vector-Valued Objective Functions, J. of Optim. Theory and Appl., Vol. 9, No. 5, (1972).

102. Lebeder, B.D., et al., Optimization Problems for an Ordered Set of Criteria, Econ. and Math. Methods, Vol. 7, No. 4, (1971).

103. Moiseev, N.N., Computational Problems in the Theory of Hierarchical Control Systems, in "Proc. 5th All-Union Conf. on Control, Part 1." Nauka, (1971).
104. Larichev, O.I., Man-Machine Solution Concepts (A Survey), ABT, No. 12, (1971).
105. Emelyanov, S.V., et al., Problems and Methods of Solution Concepts (A Survey), Int'l. Center for Sci. and Tech. Info., Moscow, (1973).
106. Diner, I.Ya., The Division of a Set of Vectors into Characteristic States and the Problem of Choosing a Solution, Problems of General and Social Prediction, No. 2, Operations Research, Information Bulletin, No. 14(29), Moscow, (1969).
107. Diner, I.Ya., The Division of a Set of Vectors into Characteristic States and the Problems of Choosing a Solution, in "Operations Research: Methodological Aspects." Nauka, (1972).
108. Germeier, Yu.B., "Methodological and Mathematical Foundations of Operations Research and Game Theory." Izd. MGU, (1967).
109. Germeier, Yu.B., Multicriteria Optimization Problems and the Role of Information, in "Proc. 5th All-Union Conf. on Control, Part 1." Nauka, (1971).
110. Germeier, Yu.B., Discussion in "Operations Research: Methodological Aspects." Nauka, (1972).
111. Emelyanov, S.V., et al., Problems of Solution Concepts in Organizational Systems, in "Proc. 5th All-Union Conf. on Control, Part 1." Nauka, (1971).
112. Vilkas, E.I., and Maiminas, E.Z., On the Problem of the Complexity of a Solution, Kybernetika, No. 5, (1968).
113. Vilkas, E.I., Utility Theory and the Concept of a Solution, in "Mathematical Methods in the Social Sciences, Seminar Proc., Vol. 1." Vilnius, (1972).
114. Yudin, D.B., A New Approach to the Formalization of the Choice of a Solution in Complex Situations, in "Proc. 5th All-Union Conf. on Control, Part 1." Nauka, (1971).
115. Ozernoi, V.M., The Concept of a Solution, ABT, No. 11, (1971).
116. Krasnenker, A.S., On the Behavior of Systems with Several Criteria, in "Proc. Math. Faculty of Voronezh University, Voronezh, (1971).
117. Krasnenker, A.S., On an Adaptive Approach to the Problem of the Concept of a Solution with Respect to Several Criteria, in "Questions of Optimal Programming in Industrial Problems." Voronezh University, Voronezh, (1972).
118. Volkovich, V.L., Multicriteria Problems and Methods for their Solutions, in "Complex Control Systems." Izd. Naukova Dumka, Kiev, (1969).
119. Alexandrov, A.A., et al., Quantitative Comparisons of Alternative Variants or Strategies in Control Problems, in "Proc. 5th All-Union Conf. on Control, Part 1." Nauka, (1971).

120. Teiman, A.I., On the Form of the Criterion in the Control of Large Systems, in Proc. 5th All-Union Conf. on Control, Part 1." Nauka, (1971).
121. Voronov, A.A., and Teiman, A.I., On Several Problems Arising in the Concept of a Solution in the Control of Large Systems under Nonnegativity Conditions, in "Proc. of the Jubilee Session of the General Section on Mech. and Control." Acad. Sci. USSR, Moscow, (1971).
122. Pervozvanskii, A.A., et al., The Control of Technological Complexes under Nonnegativity Conditions, in "Proc. 5th All-Union Conf. on Control, Part 1." Nauka, (1971).
123. Makarov, I.M., et al., The Concept of a Solution in the Choice of a Variant of a Complex Control System, ABT, No. 3, (1971).
124. Kafarov, V.V., et al., Solution Methods for Multicriteria Control Problems in Complex Chemical Engineering Systems, Dok. Acad. Sci. USSR, Vol. 198, No. 1, (1971).
125. Emelyanov, S.V., et al., Models and Methods of Vector Optimization, Summary of Science and Technology, TK, Vol. 5, (1973).
126. Emelyanov, S.V., et al., Preparations and the Concept of a Solution in Constrained Control Systems, Summary of Science and Technology, (1971).
127. Tokarev, V.V., Optimization of Parameters of a Dynamical System and Universals for Maneuvering under Different Degrees of Information, ABT, No. 8, (1971).
128. Razumikhin, B.S., Methods of Physical Modeling in Mathematical Programming and Economics, I-V, ABT, Nos. 3, 4, 6, 11, (1972); No. 2, (1973).
129. Gorochovik, V.V., On a Problem of Vector Optimization, Izv. Acad. Sci. USSR, No. 6, (1972).
130. Kozhinskaya, L.I., and Slutskii, L.I., The Role of the Contraction Approach in Vector Optimization, ABT, No. 3, (1973).
131. Gus'kov, Yu.P., Optimization of Discrete Stochastic Systems under Two Criteria, ABT, No. 10, (1970).
132. Borisov, V.I., Some Problems of Vector Optimization, in Proc. 5th All-Union Conf. on Control, Part 1." Nauka, (1971).
133. Borisov, V.I., The Choice of a Solution in the Case of Several Criteria, Problems of General and Social Prediction, Info. Bull., No. 29, Moscow, (1969).
134. Borisov, V.I., Problems of Vector Optimization, in "Operations Research: Methodological Aspects." Nauka, (1972).
135. Borisov, V.I., Discussion in "Operations Research: Methodological Aspects." Nauka, (1972).
136. Gutkin, L.S., On the Synthesis of Systems with Unconditional Preference Criteria, Izv. Acad. Sci. USSR, No. 3, (1972).
137. Gutkin, L.S., On the Synthesis of Radio Systems with Several Criteria, Radiotechnology, No. 7, (1972).
138. Gutkin, L.S., On the Application of the Method of Extreme Points for the Synthesis of Systems with a Vector Criterion, I, Izv. Acad. Sci. USSR, No. 4, (1973).

139. Krizhanovskii, G.A., Complex Criteria for Optimal Engineering Instruments, Measurement Techniques, No. 3, (1971).
140. Yuttler, Ch., Linear Models with Several Criterion Functions, Economics and Math. Methods, Vol. 3, No. 3, (1967).
141. Germeier, Yu.B., On the Contraction of Vector Criteria to a Single Criterion in the Presence of Nonnegative Contraction Parameters, Energy, Vol. 6, (1971).
142. Salukvadze, M.E., On the Optimization of Vector Functionals, 1. Programming Optimal Trajectories, ABT, No. 8, (1971).
143. Salukvadze, M.E., On the Optimization of Vector Functionals, 2. Analytic Construction of Optimal Regulators, ABT, No. 9, (1971).
144. Salukvadze, M.E., On the Optimization of Control Systems with Vector Criteria, in "Proc. 5th All-Union Conf. on Control, Part 2." Nauka, (1971).
145. Salukvadze, M.E., On Linear Programming Problems with Vector Criteria, ABT, No. 5, (1972).
146. Salukvadze, M.E., On Optimization of Control Systems According to Vector-Valued Performance Criteria, in "IFAC 5th World Congress Reports." Paris (1972).
147. Salukvadze, M.E., and Metreveli, D.G., On Problems of Optimal Flight to a Given Point, in "Automatic Control" Vols. 11-12, section 1. Metsniereba, Tbilisi, (1973).
148. Salukvadze, M.E., Vector Functionals in Linear Problems of Analytic Construction, ABT, No. 7, (1973).
149. Salukvadze, M.E., On the Existence of Solutions in Problems of Optimization under Vector-Valued Criteria, J. of Optim. Theory and Appl., Vol. 13, No. 2, (1974).
150. Athans, M., and Falb, P.L., "Optimal Control." McGraw-Hill, New York, (1966).
151. Leitmann, G., Sufficiency Theorems for Optimal Control, J. of Optim. Theory and Appl., Vol 2, No. 5, (1968).
152. Leitmann, G., A Note on a Sufficiency Theorem for Optimal Control, J. of Optim. Theory and Appl., Vol. 3, No. 1, (1969).
153. Leitmann, G., and Stalford, H., A Sufficiency Theorem for Optimal Control, J. of Optim. Theory and Appl., Vol. 8, No. 3, (1971).
*154. Miele, A., "Mechanics of Flight." Nauka, (1965).
155. Aizerman, M.A., Lectures on the Theory of Automatic Control, Goz. Fiz. Mat., (1958).
156. Voronov, A.A., "Elements of Automatic Control Theory." Voenizdat, (1954).
157. Besekerskii, V.A., and Popov, E.P., "The Theory of Automatic Control Systems." Nauka, (1972).
158. Technical Cybernetics. The Theory of Automatic Control, in "Machine Construction" (V.V. Solodovnikova, ed.), Vol. I-III. (1967-69).
159. Malkin, I.G., "Stability of Motion." Nauka, (1966).
160. Kirillova, F.M., On Problems of Analytic Construction of Regulators, PMM, Vol. 25, No. 3, (1961).

161. Kurzweil, Ya., On the Analytic Construction of Regulators, ABT, No. 6, (1961).
162. Salukvadze, M.E., Analytic Construction of Regulators. Constraints of Perturbed Motion, ABT, No. 10, (1961).
163. Salukvadze, M.E., On the Analytic Construction of Optimal Regulators under Constantly Acting Disturbances, ABT, No. 6, (1962).
164. Salukvadze, M.E., On the Question of Analytic Construction of Optimal Regulators, ABT, No. 4, (1963).
165. Salukvadze, M.E., On the Question of Invariance of Optimal Regulators, ABT, No. 5, (1964).
166. Salukvadze, M.E., On the Problem of Synthesis of an Optimal Regulator in Linear Systems with a Delay subject to Constantly Acting Disturbances, ABT, No. 12, (1962).
167. Salukvadze, M.E., On the Synthesis of Optimal Regulators in Systems with a Delay, Proc. Georgian Acad. Sci. SSR, Vol. 37, No. 1, (1965).
168. Salukvadze, M.E., The Analytic Construction of Optimal Regulators in Systems with a Delay, Proc. Georgian Acad. Sci. SSR, Vol. 38, No. 3, (1966).
169. Salukvadze, M.E., Optimal Search for the Extrema of a Function of Many Variables, in "Automatic Control. Metsniereba, Tbilisi, (1967).
170. Kantorovich, L.V., "Mathematical Methods in Constrained and Planned Production." LGU, (1939).
171. Kantorovich, L.V., On an Effective Solution Method for some Classes of Extremal Problems, Pok. Acad. Sci. SSR, Vol. 28, No. 3, (1940).
*172. Gass, S., "Linear Programming." Fizmatgiz, (1961).
173. Zukovitskii, S.I., and Avdeeva, L.I., "Linear Programming." Nauka, (1967).
174. Vil'chevskii, N.O., and Razumikhin, B.S., Mechanical Models and Solution Methods for General Linear Programming Problems, ABT, No. 4, (1966).
175. Fel'dbaum, A.A., "Computational Methods in Automatic Systems." Fizmatgiz, (1959).
176. Rastrigin, L.A., "Statistical Search Methods." Fizmatgiz, (1968).
*177. Wilde, D.J., "Optimum Seeking Methods." Fizmatgiz, (1967).
178. Ribashov, M.V., Gradient Methods for the Solution of Convex Programming Problems with an Electronic Model, ABT, No. 11, (1965).
179. Ribashov, M.V., Gradient Methods for the Solution of Linear and Quadratic Programming Problems with an Electronic Model, ABT, No. 12, (1965).
180. Benaiun, R., et al., Multiple Criteria Linear Programming. The Method of Constraints, ABT, No. 8, (1971).
181. Er'moleva, L.G., On a Solution Method for Linear Vector Optimization Problems, in "Math Methods and Systems Optimization, Kiev, (1971).

182. Kuz'min, I.V., et al., Methods for Constructing Global Criteria in Mathematical Programming Problems, Mech. and Auto. Control, No. 6, (1971).
183. Zak, Yu.A., Models and Methods for the Construction of Compromise Plans in Mathematical Programming Problems with Several Objectives, Kybernetika, No. 4, (1972).
184. Volkovich, V.L., and Darienko, L.F., On an Algorithm for Choosing a Compromise Solution for Linear Criteria, Kybernetika, No. 5, (1972).
185. Ostapov, S.S., On Multicriteria Linear Progamming Problems, in "Mathematical Questions in the Formation of Economic Models." Novosibirsk, (1970).
186. Krasin, A.K., Rapid Gas-Cooling of Reactors by Gaseous Dissociation, in "Proc. Conf. of MAGATE Experts." Minsk, (July 24-28, 1972).
187. Dzhibladze, N.I., et al., Optimization of Cooling Parameters with Water Cooling, Izv. Acad. Sci. Byel. SSR, No. 4, Minsk, (1970).
188. Krasin, A.K., et al., Physical-Technical Foundations for the Construction of Nuclear Power Stations with Gas-Cooled Nuclear Reactors by Rapid Neutrons with Dissociated Heat Transfer by Tetranitrogen Oxide, World Studies on Atomic Energy, Vol. 5, (1972).
189. Nestevenko, V.B., The Physical-Technical Characteristics of Dissociated Tetranitrogen Oxide in AES with Rapid Gas-Cooled Reactors, in "Proc. Conf. of MAGATE Experts." Minsk, (July 24-28, 1972).

SUPPLEMENTAL REFERENCE LIST

(Translator's Note: The references that follow were kindly supplied by Professor George Leitmann, University of California, Berkeley, and did not appear in the original Russian work. They are meant to bring the interested reader up-to-date with recent Western literature on vector-valued optimization problems.)

1. Aubin, J.-P, A Pareto Minimum Principle, in "Differential Games and Related Topics" (H.W. Kuhn, G.P. Szego, eds.). North-Holland Publishing Co., Amsterdam, (1971).

2. Sinha, N.K., Temple, V., and Rey, A.J.C., A Vector Cost Function for Efficient Adaptation, Int'l. J. of Control, Vol. 16, No. 6.
3. Athans, M., and Geering, H., Necessary and Sufficient Conditions for Differentiable Nonscalar-Valued Functions to Attain Extrema, IEEE Trans. on Auto. Control, Vol. AC-18, No. 2, (1973).
4. Smale, S., Optimizing Several Functions, presented at the Int'l. Symp. on Manifolds, Tokyo, (April, 1973).
5. Stadler, W., Natural Structural Shapes (The Static Case), Quarterly J. of Mech. and Appl. Math., Vol. XXXI, pt. 2, (1978).
6. Borwein, J., Proper Efficient Points for Maximizations with Respect to Cones, SIAM J. of Control, Vol. 15, No. 1, (1977).
7. Letov, A, and Rozanov, Yu., On Optimal Compromise for Multidimensional Resource Distribution, IIASA Research Report 74-8, (June, 1974).
8. Schmitendorf, W., and Moriarty, G., A Sufficient Condition for Coalitive Pareto-Optimal Solutions, JOTA, Vol. 18, No. 1, (1976).
9. Shen, H, and Line, J., A Computational Method for Linear Multiple-Objective Optimization Problems, in "Proc. 1975 Conf. on Decision and Control."
10. Zeleny, M., (ed), "Multiple Criteria Decision Making." Lecture Notes in Economics and Math. Systems, Springer Verlag, Berlin, (1976).
11. Yu, P.L., and Leitmann, G., Nondominated Decisions and Cone Convexity in Dynamic Multicriteria Decision Problems, JOTA, Vol. 14, No. 5, (1974).
12. Tamura, K., A Method for Constructing the Polar Cone of a Polyhedral Cone, with Applications to Linear Multicriteria Decision Problems, JOTA, Vol. 19, No. 4, (1976).
13. Leitmann, G., Some Problems of Scalar and Vector-Vlaued Optimization in Linear Viscoelasticity, JOTA, Vol. 23, No. 1, (1977).
14. Naccache, P., Connectedness of the Set of Nondominated Outcomes in Multicriteria Optimization, JOTA, Vol. 25, No. 3, (1978).
15. Stadler, W., A Survey of Multicriteria Optimization or the Vector Maximum Problem, Part I: 1776-1960, to appear in JOTA.
16. Yu, P.L., Decision Dynamics with an Application to Persuasion and Negotiation, in "TIMS Studies in Management Science," pp. 159-65. North-Holland Publishing Co., Amsterdam, (1975).
17. Leitmann, G., and Marzollo, A., (eds.), "Multicriteria Decision Making." Springer Verlag, Wien-New York, (1975).
18. Leitmann, G., (ed.), "Multicriteria Decision Making and Differential Games." Plenum Press, New York, (1976).